Proceedings

Ein stetig steigender Fundus an Informationen ist heute notwendig, um die immer komplexer werdende Technik heutiger Kraftfahrzeuge zu verstehen. Funktionen, Arbeitsweise, Komponenten und Systeme entwickeln sich rasant. In immer schnelleren Zyklen verbreitet sich aktuelles Wissen gerade aus Konferenzen, Tagungen und Symposien in die Fachwelt. Den raschen Zugriff auf diese Informationen bietet diese Reihe Proceedings, die sich zur Aufgabe gestellt hat, das zum Verständnis topaktueller Technik rund um das Automobil erforderliche spezielle Wissen in der Systematik aus Konferenzen und Tagungen zusammen zu stellen und als Buch in Springer.com wie auch elektronisch in SpringerLink und Springer Professional bereit zu stellen.

Die Reihe wendet sich an Fahrzeug- und Motoreningenieure sowie Studierende, die aktuelles Fachwissen im Zusammenhang mit Fragestellungen ihres Arbeitsfeldes suchen. Professoren und Dozenten an Universitäten und Hochschulen mit Schwerpunkt Kraftfahrzeug- und Motorentechnik finden hier die Zusammenstellung von Veranstaltungen, die sie selber nicht besuchen konnten. Gutachtern, Forschern und Entwicklungsingenieuren in der Automobil- und Zulieferindustrie sowie Dienstleistern können die Proceedings wertvolle Antworten auf topaktuelle Fragen geben.

Today, a steadily growing store of information is called for in order to understand the increasingly complex technologies used in modern automobiles. Functions, modes of operation, components and systems are rapidly evolving, while at the same time the latest expertise is disseminated directly from conferences, congresses and symposia to the professional world in ever-faster cycles. This series of proceedings offers rapid access to this information, gathering the specific knowledge needed to keep up with cutting-edge advances in automotive technologies, employing the same systematic approach used at conferences and congresses and presenting it in print (available at Springer.com) and electronic (at SpringerLink and Springer Professional) formats.

The series addresses the needs of automotive engineers, motor design engineers and students looking for the latest expertise in connection with key questions in their field, while professors and instructors working in the areas of automotive and motor design engineering will also find summaries of industry events they weren't able to attend. The proceedings also offer valuable answers to the topical questions that concern assessors, researchers and developmental engineers in the automotive and supplier industry, as well as service providers.

Weitere Bände in der Reihe http://www.springer.com/series/13360

Rolf Isermann
(Hrsg.)

Fahrerassistenzsysteme 2016

Von der Assistenz zum automatisierten Fahren
2. Internationale ATZ-Fachtagung

Springer Vieweg

Herausgeber
Rolf Isermann
Technische Universität Darmstadt
Darmstadt, Deutschland

ISSN 2198-7432   ISSN 2198-7440   (electronic)
Proceedings
ISBN 978-3-658-21443-2   ISBN 978-3-658-21444-9   (eBook)
https://doi.org/10.1007/978-3-658-21444-9

Die Deutsche Nationalbibliothek verzeichnet diese Publikation in der Deutschen Nationalbibliografie; detaillierte bibliografische Daten sind im Internet über http://dnb.d-nb.de abrufbar.

Vorwort

Die zweite ATZlive-Fachtagung „Fahrerassistenzsysteme" 2016 bestand aus 5 Übersichtsbeiträgen und 7 Vortragsbereichen zu den Themen: Märkte und Nutzer, Sensorik, Safety und Security, Simulation und Test, Komponenten, E/E-Architekturen und Nutzfahrzeuge.

Es wurden z. B. folgende *allgemeine Aussagen* zu derzeitigen und zukünftigen Entwicklungen gemacht:

- Wichtige Trends sind: Demographische Entwicklung, Urbanisierung, Energieknappheit, Klimaschutz, Internet der Dinge (auch Fahrzeuge und ihre Komponenten).
- Unfälle sind seit 1992 rückläufig. Seit 2010 zeigt sich eine Stagnation. Eventuell hilft hier das automatische Fahren weiter.
- Das automatische Fahren wird eine evolutionäre Entwicklung von niederer Geschwindigkeit zu höherer Geschwindigkeit zeigen.
- Wesentlich ist eine hervorragende Umgebungserfassung mit viel Redundanz
- Das Highly Automated Driving (HAD) setzt für alle Systeme ein fail operational Verhalten voraus, d. h. Redundanz im Fehlerfall. Dies betrifft Bremsen, Lenkungen, Spannungsquellen und eine fail-safe elektronische Architektur.
- Es werden digitale Straßenkarten mit hoher Auflösung notwendig.
- Eine Vernetzung von Daten aus dem Fahrzeug mit in der Cloud (Backend) gespeicherten Informationen wird sich entwickeln.
- Es sollte eine möglichst unabhängige Cloud-Plattform für alle angestrebt werden.
- Die neue Mercedes-Benz E-Klasse hat ein Fahrerassistenzpaket der Generation 4. Es enthält eine verbesserte Umgebungserfassung, erlaubt teilautomatisiertes Fahren und neue Funktionalitäten, wie z. B. den Spurwechselassistent, bei dem der Fahrer unterstützt wird, und automatisches Bremsen bei in eine Kreuzung eindringenden Fahrzeugen. Stereo-Kamera und mehrere Radarsensoren erlauben eine dreidimensionale Fusion. Beim automatischen Fahren können die Hände 30 s lang vom Lenkrad weg bleiben. Dann erfolgt ein erstes akustisches Warnzeichen.
- Volvo kündigt für 2020 an: Keine Unfälle, keine getöteten Personen oder Schwerverletzten. Drive Me ist ein „Self-Driving Laboratory" mit 100 teilautomatisch fahrenden Fahrzeugen in den Jahren 2017 bis 2020 in der Umgebung von Göteborg und in China 2018. Fahrzeuge sind XC90 und XC60. Mehr als 20 Sensoren erfassen die Umgebung und die Übergabezeiten werden mit 10 bis 30 s angegeben.
- Das vollautomatische Fahren wird in 15 bis 20 Jahren erwartet.

Die in Sitzungen gegliederten Vortragsbereiche enthielten z. B. ausschnittsweise folgende Feststellungen:

Märkte, Nutzer, Sensorik

- Es wird erwartet, dass die Unfallzahlen durch ADAS reduziert werden, was auch die Reparaturkosten beeinflusst.
- AEB (Automatic Emergency Braking) wird zunächst für auffahrende Fahrzeuge entwickelt, dann für Fußgänger und Radfahrer.
- 50% der Unfälle mit getöteten Fußgängern geschehen nachts.
- Das Abkommen von der Fahrbahn erfolgt zu 75% auf gerader Straße.
- Das automatische Parken muss innerhalb von Parkhäusern ohne GPS auskommen. Eine Bluetooth Ortung mit Funkstellen (Beacons) ist nur 50 cm genau.

Safety und Security

- Eine Gefahr für Fahrzeuge wird in zwei bis drei Jahren erwartet. Bzgl. des Angriffsschutzes sollte man aus der IT Branche lernen.
- Potentielle Eintrittsstellen sind zzt. OBD-Stecker, Bluetooth-Verbindungen, Keyless Entry.
- Die Connectivity erfordert eine ECU-Struktur, die vertrauenswürdig ist. Hier bieten sich 4-Ebenen-Schalenmodelle an: ECU, Secure Communication, Security Gateways und Firewalls.
- Das Automotive Security Engineering schließt auch eine Security Maintenance ein, die schließlich ein Firmware Over The Air (FOTA) Update erforderlich macht. Hierdurch entstehen zusätzliche Kosten.
- Eine Podiumsdiskussion unterstrich die Bedeutung der Robustheit gegenüber Hackerangriffen (Resilience). Die OEMs sind hier aktiv; die Weiterentwicklung geschieht inkrementell. Die als Beispiel aus den USA gezeigten Hacker-Eingriffe hatten Fahrzeuge mit relativ alten Elektronikarchitekturen. In der Luftfahrt zeigt sich eine ähnliche Situation.
- Security updates über Steckverbindungen und Kabel kosten zu viel Zeit und sind unhandlich. Deshalb werden Software Over The Air Updates (SOTA) angestrebt.

Simulation und Test

- BMW, Daimler und Volkswagen arbeiten an einer multifunktionalen offenen Simulationsplattform für die Entwicklung und die Validierung von automatischen Fahrfunktionen.
- Testumgebungen für die virtuelle Absicherung von sicherheitskritischen Assistenzsystemen erleichtern die Entwicklung und Validierung, da Tests mit echten Fahrzeugen nicht ausreichend sein werden. Hardware-in-the-Loop Systeme (HiL), aber auch MiL und SiL erlauben die Entwicklung und Validierung von Sicherheitsfunktionen. Dabei wird die entsprechende Software vor den echten Fahrzeugtests geprüft.
- Hochautomatisierte Fahrzeuge benötigen außer einer Überwachung aller aktiven Systeme eine fehlertolerante Auslegung der sicherheitsrelevanten Komponenten, wie z. B. Bremsen, Lenkungen, Energieversorgung und Kommunikationswege. Das Ziel ist hierbei ein Fail-Operational-Verhalten nach Auftreten eines Fehlers zu erhalten. Dies kann durch Hardware- und teilweise auch analytische Redundanz erfolgen. Somit müssen fehlertolerante elektrische und hydraulische Aktoren und fehlertolerante Sensorsysteme entwickelt werden. Ein Beispiel ist die fehlertolerante Auslegung elektrischer Servolenksysteme.
- Die E/E-Architekturen von Nutzfahrzeugen müssen ebenfalls Fail-Operational ausgelegt werden mit entsprechenden Redundanzen, besonders für Energieversorgung, Brems- und Lenksysteme.

Die Tagung erbrachte somit wichtige generelle Ergebnisse, die in den einzelnen Vortragsmanuskripten oder Vortragsfolien weiter vertieft sind.

Dieser Band enthält Beiträge der Tagung, die zur Veröffentlichung freigegeben wurden.

Prof. Dr.-Ing. Dr. h.c. Rolf Isermann
Technische Universität Darmstadt
Forschungsgruppe Regelungstechnik und Prozessautomatisierung

Inhalt

Autorenverzeichnis

Jörg Bakker Daimler AG, Stuttgart, Deutschland

Gerald-Alexander Beese KTI GmbH & Co. KG, Lohfelden, Deutschland

Michael Beine dSpace GmbH, Paderborn, Deutschland

Linus Bundschuh MAN Truck & Bus AG, München, Deutschland

Jan Dobberstein Daimler AG, Stuttgart, Deutschland

Dr. Björn Giesler Elektrobit Automotive GmbH, Erlangen, Deutschland

Dr. Karl-Heinz Glander ZF TRW – TRW Automotive, Düsseldorf, Deutschland

Bejamin Glas ETAS GmbH, Stuttgart, Deutschland

Thomas Helmer BMW Group, München, Deutschland

Heiko Herchet EDAG Engineering GmbH, Fulda, Deutschland

Dr. Claudia Hollmann dSpace GmbH, Paderborn, Deutschland

Gregor Hordys dSpace GmbH, Paderborn, Deutschland

Markus Ihle ETAS GmbH, Stuttgart, Deutschland

Prof. Dr. Dr. Rolf Isermann TU Darmstadt, Darmstadt, Deutschland

Olaf Jung BMW Group, München, Deutschland

Benjamin Jüstel Continental Teves AG & Co. oHG, Frankfurt, Deutschland

Ronald Kates REK Consulting, Otterfing, Deutschland

Dr. Roman Katz Ibeo Automotive Systems GmbH, Hamburg, Deutschland

Helge Kiebach KTI GmbH & Co. KG, Lohfelden, Deutschland

Dr. Markus Koegel ESCRYPT GmbH, Bochum, Deutschland

Dr. Andreas Lapp Robert Bosch GmbH, Stuttgart, Deutschland

Angelo Raczek Adam Opel AG, Rüsselsheim, Deutschland

Andreas Rigling ADAC e.V. Technik Zentrum, Landsberg, Deutschland

Andre Rolfsmeier dSpace GmbH, Paderborn, Deutschland

Lex van Rooij ZF TRW – TRW Automotive, Düsseldorf, Deutschland

Dr. Georg Niedrist TTTech Automotive GmbH, Wien, Österreich

Elina Schäfer EDAG Engineering GmbH, Fulda, Deutschland

Michael Schaffert Robert Bosch GmbH, Stuttgart, Deutschland

Markus Schöttle Springer Fachmedien Wiesbaden GmbH, Wiesbaden, Deutschland

Gerhard Steiger Bosch Chassis Systems Control Division, Heilbronn, Deutschland

Ulrich Stöckmann Continental Teves AG & Co. oHG, Frankfurt, Deutschland

Alexander Süssemilch EDAG Engineering GmbH, Fulda, Deutschland

Thomas Thiel Robert Bosch GmbH, Stuttgart, Deutschland

Timo Vogt BMW Group, München, Deutschland

Andreas Wagener IPG Automotive GmbH, Karlsruhe, Deutschland

Lei Wang BMW Group, München, Deutschland

Dr. Alexander Weitzel Adam Opel AG, Rüsselsheim, Deutschland

Gernot Wiese Adam Opel AG, Rüsselsheim, Deutschland

Dr. Marko Wolf ESCRYPT GmbH, Bochum, Deutschland

Albert Zaindl MAN Truck & Bus AG, München, Deutschland

Towards a new mobility – the driver becomes the passenger

Gerhard Steiger, President Bosch Chassis Systems Control Division

© Springer Fachmedien Wiesbaden GmbH, ein Teil von Springer Nature 2018
R. Isermann (Hrsg.), *Fahrerassistenzsysteme 2016*, Proceedings,
https://doi.org/10.1007/978-3-658-21444-9_1

Introduction

Mobility patterns are changing. The trends we perceive indicate that a completely new understanding of mobility is coming up. The role of the driver within individual mobility is particularly affected. We can imagine that in the future, in certain situations, the car will take over the "driving seat" and the driver will enjoy the ride as a passenger. This is the vision of automated driving that Bosch aims to make a reality.

How mature are the approaches that lead us to this new form of mobility? How soon can we expect to buy a highly automated car to drive along our public roads? What specific challenges are we currently facing? Which issues must be solved and which tasks completed to make highly automated driving become true?

This lecture will provide some answers to these questions. In summary: automated driving shows the potential to provide remarkable benefits to individual mobility. This expectation generates the momentum we can already observe following the public discussions in several countries and justifies the investment in the related technologies. On the other hand, there is still a long way to go until all technical and regulatory challenges are resolved.

Expected benefits

Multiple indicators related to aging society, urbanization, fossil-fuel depletion and global warming, as well as increasingly ubiquitous connectivity, prefigure some radical changes in different regional societies and markets: the demographic segment above 65 years is growing twice as fast as other segments. By 2050, the world population will have risen from about 7 billion today to about 9 billion people. In 2030, about 60 percent of the world population – roughly 5 billion people – will live in cities. 50 billion things will be networked by 2020. According to the latest forecasts, global warming is unlikely to be kept below 2 degrees.

Worldwide, more and more people will aim to become mobile. A few specific examples might be brought up to underline this statement: worldwide, 1.2 billion people spend an average of 50 minutes per day in the car, often in traffic jams, according to a McKinsey study. ADAC announced that 2014 was the record year for traffic jams in Germany. The management consultants of Roland Berger revealed that every driver wastes 100 hours per year searching parking lots for spaces.

These trends are obviously having an impact on the automotive world. Reading newspapers or visiting the trade fairs, it becomes apparent that the future of mobility relies on connectivity, automation and electrification of vehicles.

As a matter of fact, we forecast a stronger integration of the home entertainment products in vehicles: today, drivers also expect the newest software and the latest apps in the car. All vehicles will be connected to each other and with the traffic infrastructure via a cloud.

This decade, a clear reduction in battery costs is expected which will increase the appeal of vehicles with electric drive trains. Diversified multimodal mobility will combine intelligently different mobility offers to a larger extent, for local traffic as well as for buses and railway.

Automation is a response aimed at ensuring a high quality of life, despite the significant increase of road utilization. Automated driving is therefore the word on everybody's lips. As previously mentioned, more than 9 billion people will be living on earth in 2050 according to estimates by the United Nations. However, more people also means a much higher traffic volume and with it, a growing risk of road accidents.

If we consider the worldwide traffic development over the last decade, we can see a very positive general trend. During the period 2003 to 2013, the number of the accidents with injuries in the countries represented in figure 1 reduced by approximately 15 percent overall. The positive effects of good infrastructure, road safety education and safety systems in the car are clearly visible in North America, Germany and Japan. On the other hand, in emerging regions such as India or Brazil, the development is rather negative.

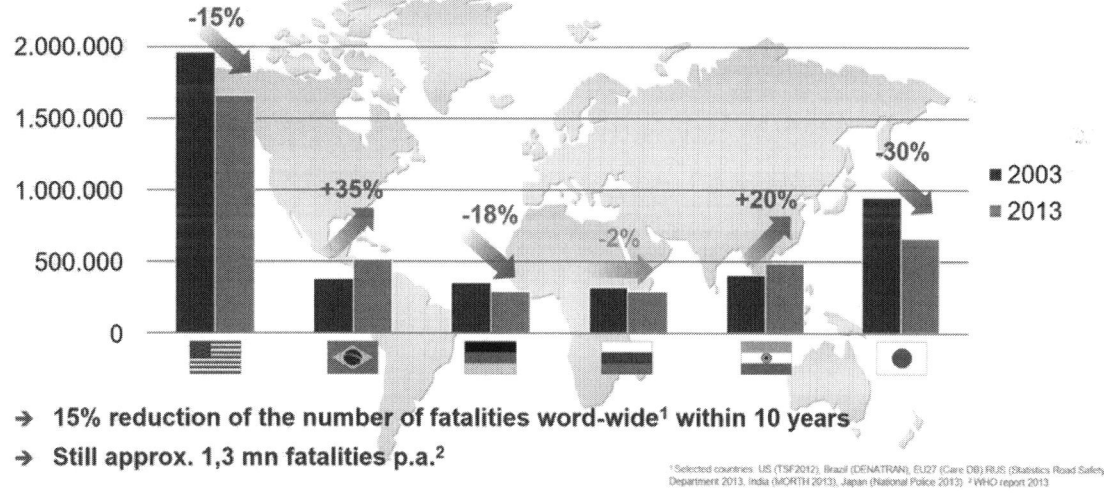

Figure 1: Road safety – More automation saves lives

However, going even deeper into the analysis of the overall road safety situation e.g. in Germany, it becomes apparent that the positive evolution of the last few decades has stopped. Figure 2 shows the proportion of severely injured versus the proportion of fatally injured road users. After a phase of impressive improvement, the figures indicate stagnation over the last 2-3 years.

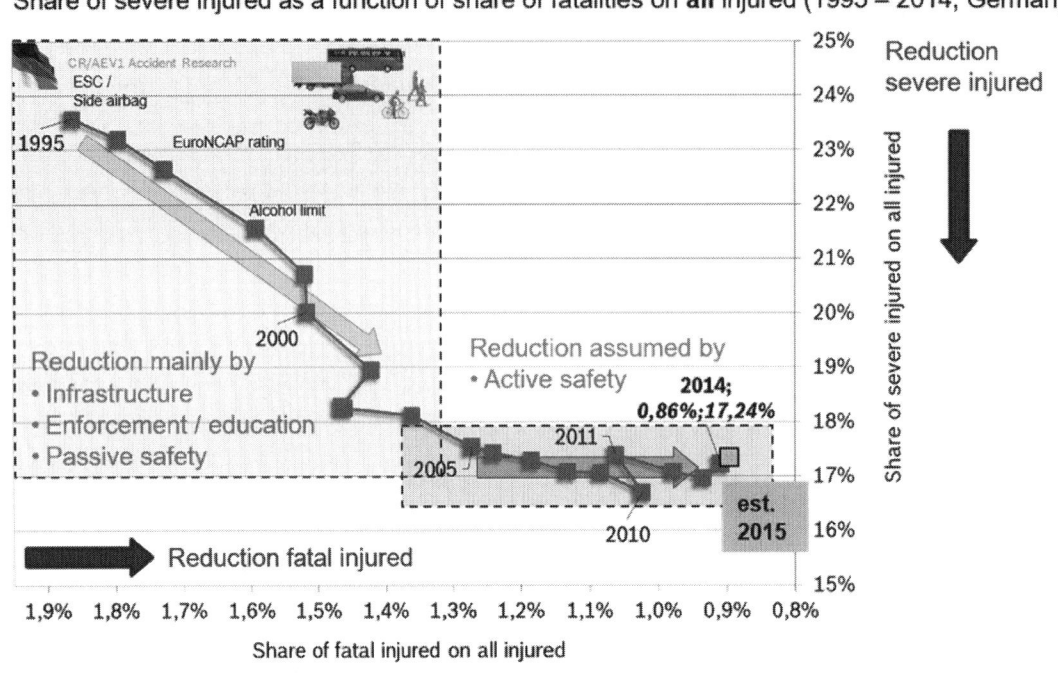

Figure 2: Traffic safety by severity since 1995 – Stagnation, no progress estimated for 2015

The Bosch accident research has analyzed the situation in Germany in detail. Since the beginning of the century, the number of the road accidents with physical injuries decreased by roughly 30 percent. Reasons for this include an improved traffic infrastructure, progress in road safety education and the improvement of the active safety systems in vehicles, such as the electronic stability program. Since its launch to the market by Bosch, this system has avoided approximately 260,000 accidents and has saved more than 8500 lives across Europe. It is worth remembering that, despite all our enthusiasm concerning automation, the installation rate of ESP® worldwide is still only around 64 percent in 2015. This means that 1/3 of the vehicles are not fulfilling a key pre-requisite for more advanced assistance functions yet.

Future driver assistance systems show great potential for preventing accidents completely, or at least for reducing the consequences of an accident. Assuming that the installation rate of driver assistance systems like automatic emergency braking, lane and intersection assist will develop at a similar rate to the installation of the electronic stability program, 45 percent of the current accidents can be addressed by systems already available on the market or currently under development (see figure 3). A further 37 percent of the current accidents could be addressed by increasing the automation of vehicles (e.g. risky overtaking manoeuvers, drowsiness etc.).

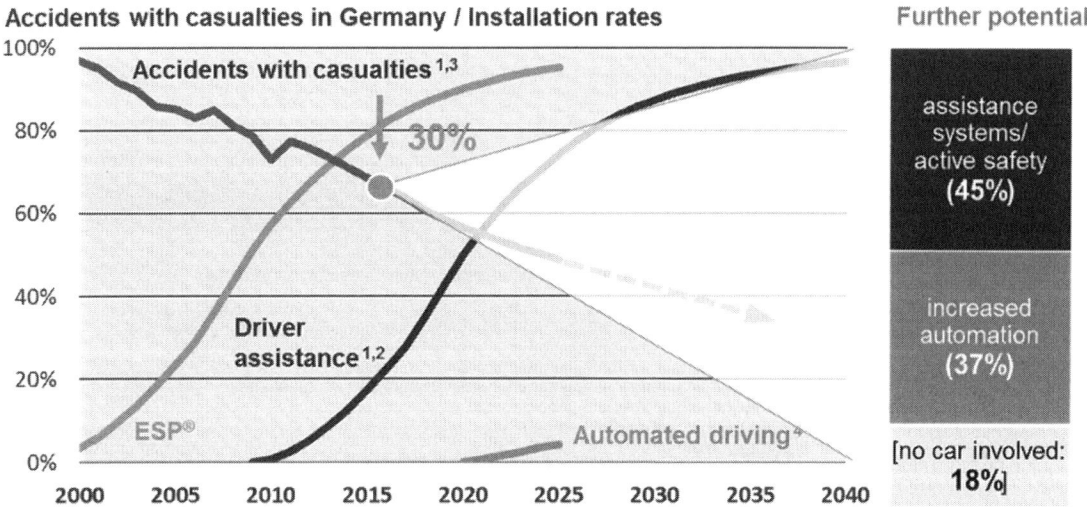

Accidents with casualties in Germany / Installation rates   Further potential:

Figure 3: Road safety – More automation saves lives. Increased automation can address 37% of today's accidents

Every year, an estimated 1.3 million people worldwide die on the road. At the same time, around 90 percent of all accidents are attributable to human error. The origin might be an overstressed or an under-stimulated driver who is easily distracted and thus makes serious errors. Automated driving functions that relieve drivers from driving tasks in confusing or monotonous situations can therefore save lives.

On the other hand, driving comfort which releases the driver from monotony can definitively increase safety. The result of a field-operational-test financed by the EU proved this insight in 2012.

Adaptive Cruise Control (ACC) was originally supposed to be a comfort function to assist drivers during longer trips on highways and country roads. ACC always keeps the set safety distance from the vehicle ahead and automatically adjusts the driving speed to the flow of traffic by accelerating and braking. In combination with a collision warning system, ACC can reduce the number of severe braking maneuvers on motorways by 67 percent. At the same time, 73 percent fewer cases of vehicles following each other within critical distances occur.

Consequently, the increasing level of automation leads to a stronger synchronization of traffic. This reduces the risk of congestion and accidents, and increases the overall energy efficiency of vehicles.

As the driver is relieved of his driving tasks, he is able to use his time for pursuing other activities on board. This will widen the moments of driving pleasure beyond the simple act of driving: the driver could enjoy the ride as an experience of productive work, media interaction or communication with the world outside (see figure 4).

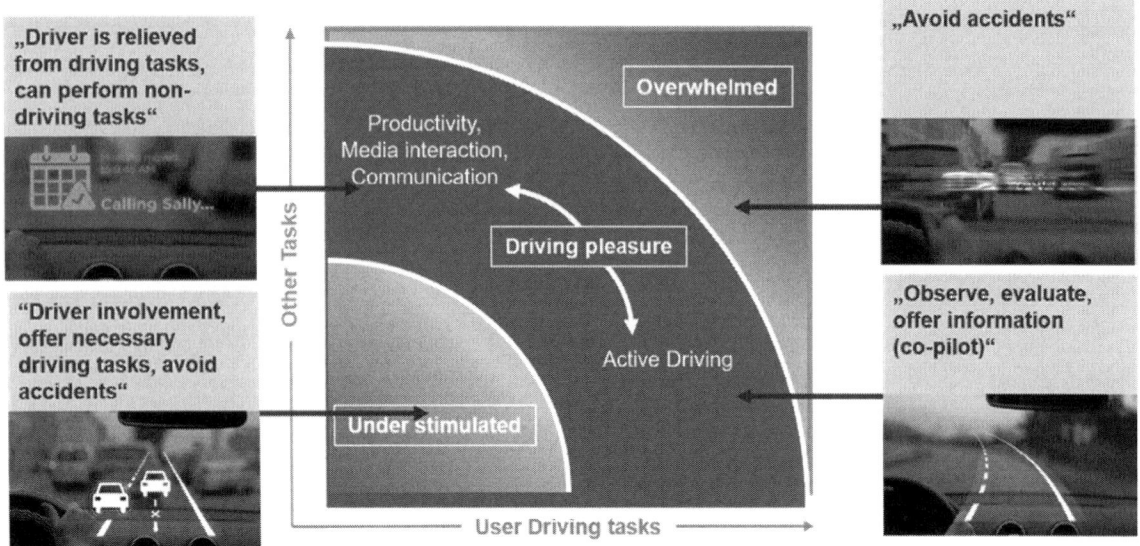

Figure 4: Cars are the better drivers – Automated driving saves lives and provides driving fun

Steps towards automation

But let's not get ahead of ourselves: it will still take some time before cars can drive fully automatically from door to door and master all situations completely by themselves during the trip. This is not to be expected before the second half of the next decade. However, the fundamental technical challenges will largely be resolved by the end of this decade.

Little by little, modern cars are thus taking over an increasing level of control. The driver's task will nevertheless change gradually at the same time, as already mentioned, and namely to the degree in which automation is progressing. Since driver assistance systems need to be constantly monitored, the responsibility remains with the driver. Highly automated functions are able to release the driver from his duties for a while; however, in a defined time period, he must always be ready to take over if required. Fully automated functions do not require driver supervision. The automation capability is easier to realize with comparable, stable or predictable situations like highway driving or parking. This is the reason why development work is currently focusing on these two topics.

Park assistance systems are already quite popular in many countries. For instance, in Germany every second car registered in 2014 was equipped with such a system, according to Bosch analysis. However, the number of parking and maneuvering accidents has increased by 30 percent over the last 10 years. Meanwhile, misfortunes such as these account for about 40 percent of all passenger car accidents resulting in property damage, according to a study published by Allianz AG in 2015.

Last year Bosch introduced a remote park assist system with which the vehicle parks completely by itself – even in the garage. The driver no longer has to sit in the vehicle while it parks, providing that he monitors the event from the outside. The function breaks off if the driver releases the control device, which might be his smartphone.

The next step is the introduction of an auto park pilot. This system parks the vehicle in parking spaces without needing to be monitored by the driver. The final stage of development is the so-called Valet Parking. Here, the driver has only to leave his vehicle in a predetermined place, e.g. at the mall's entrance and the vehicle finds its way autonomously to a nearby car park.

The path towards automated driving on freeways and country roads starts with driver assistance systems. Among these functions, the automatic emergency braking (AEB) assist has a particularly high potential for preventing accidents. If the system detects an obstacle or a person in the lane ahead, it warns the driver and prepares the braking system for full braking. If the driver does not react, the assistant automatically triggers emergency braking. In Germany alone, up to 72 percent of all rear-end collisions resulting in personal injuries could be prevented if all vehicles were equipped with such a system.

A further example for an assistance function is the evasive steering support from Bosch: as soon as the driver initiates a steering maneuver, the system automatically increases the steering torque. The maximum steering angle is thereby achieved 25 percent earlier and the obstacle cleared by up to 60 centimeters earlier, according to studies. Every second and every centimeter is crucial when it comes to preventing a collision.

Driver assistance functions support the driver today by taking over either the longitudinal or the lateral driving task in specific situations. Automated driving introduces the first driving functions that will carry out both longitudinal and lateral control tasks simultaneously and allow the driver to be absent from the active driving task for a limited amount of time. However, the driver will be responsible for permanently supervising these functions, as mentioned before.

We expect the introduction of several partially automated functions to the market, characterized by an increasing level of performance as shown in figure 5. Traffic jam assist provides longitudinal and lateral guidance up to 60 kph, while the integrated cruise assist allows the extension of its speed range to 130 kph. Finally, the highway assist is able to make a lane change automatically, but only after driver confirmation as a sign of maintaining control of the vehicle.

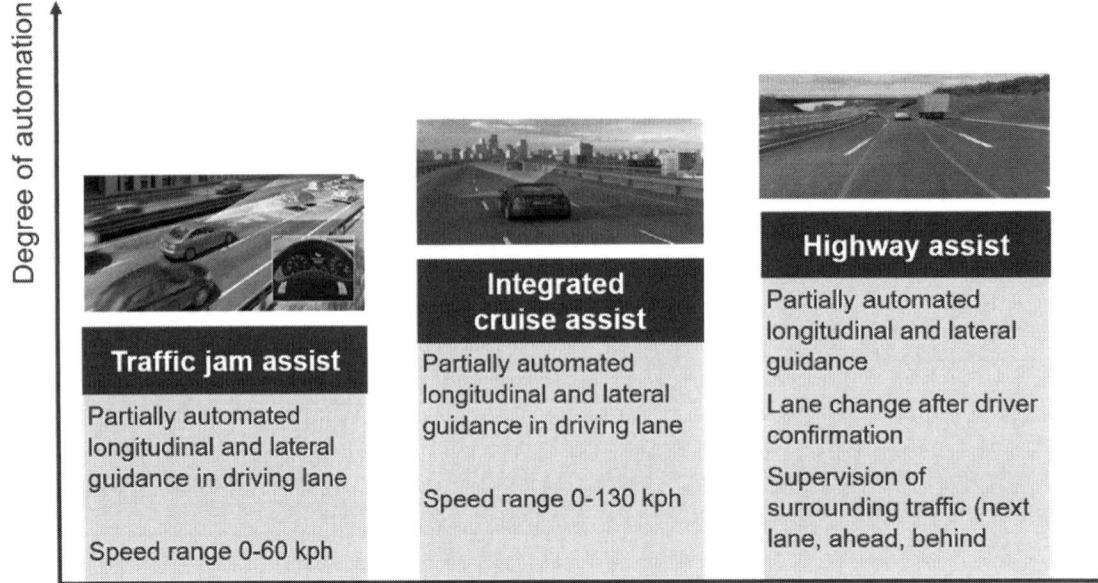

Figure 5: Development steps – automated driving; further partially automated functions will follow

Highly automated driving functions will follow soon, completing the overall picture as shown in figure 6. The first market introduction of the traffic jam pilot was already announced by an OEM as an example. We expect a wide scale introduction of such functions in well-defined situations and restricted environments first, such as highways. In this case we face unidirectional traffic flow, whereas urban driving scenarios include cross traffic situations, pedestrians and bicyclists in addition. The requirements of perception, situation recognition, and decision making that these systems must meet are considerably higher in a more complex environment.

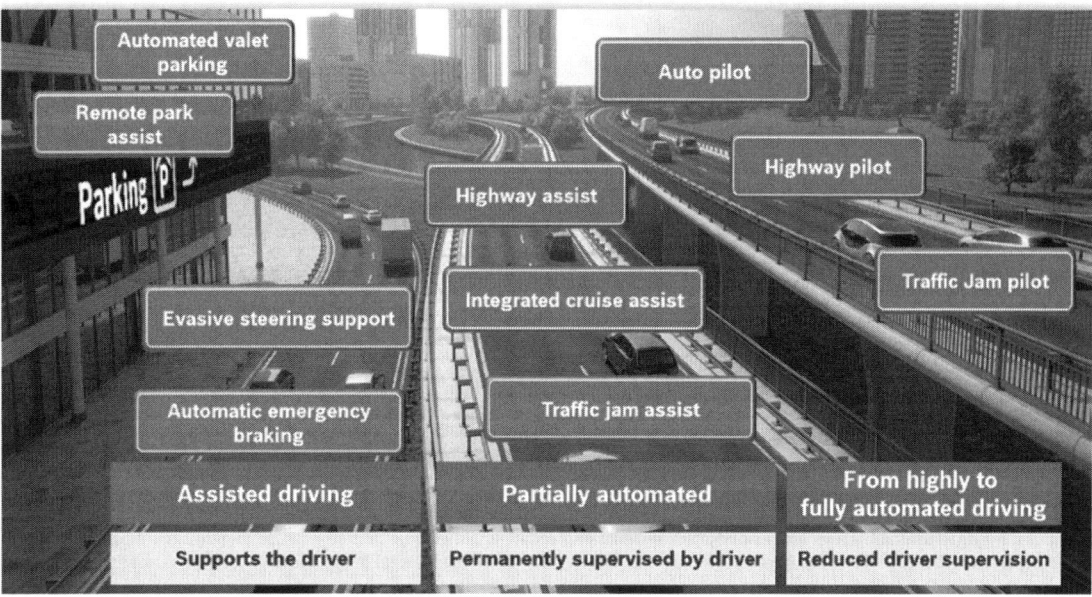

Figure 6: Level of automation – step by step approach based on a functional roadmap

Need for fail-operational architecture

Unlike driver assistance systems, partially automated driving functions take over the steering, braking and acceleration of a car in individual driving situations.

For example: The traffic jam pilot is designed to provide automated guidance of the vehicle in situations with traffic congestion on highway. This requires combined lateral and longitudinal guidance of the vehicle at velocities typically less than 60 kph on roads with more than one lane per driving direction, wide lanes and low curvature. Lateral guidance aims to keep the vehicle in the current lane; automated lane changes are not supported. Longitudinal guidance aims at keeping a safe distance to the preceding vehicle. If a system boundary is reached, the driver is requested to take over control of the vehicle. If the driver does not respond accordingly within a defined time limit, the system will start switching to the safe state.

The highway pilot will extend traffic jam pilot to higher velocities of up to 130 kph. In addition, it will provide automated lane change maneuvers and finer lateral guidance within the lane, resulting in a more comfortable distance to adjacent vehicles. A further evolution of highway pilot will include exit to exit functionality. This implements functional features such as transitions from one highway to another highway, including on-ramps and off-ramps. This allows the driver to enter a city area as a target destination in the vehicle navigation system, and the pilot function will automatically select relevant combinations of highways to reach that destination.

Highly automated driving functions cannot rely on the driver's availability for taking over the control of the vehicle in case of failure (see figure 7). Although these functions do not necessarily have to consider all situations, it is required to cover as many as possible in order to obtain higher availability of the function. A highly automated vehicle must be designed to be as safe as humanly possible. In simple terms, the safe state translates to bringing the vehicle to a standstill in a safe location while sending out warning signals to other traffic participants, e.g. via hazard messages or activated warning lights.

Figure 7: Cars take over responsibility during highly automated driving – in the case of a failure, the system maintains a degraded mode of operation

Similarly, in the case of hardware failures, the system needs to stay operational, at least with reduced functionality, e.g. until the driver has taken back control. This imposes additional requirements on the sensor set, electronic control units (ECUs), communication network, power supply and actuators. The use cases to be covered by a respective function typically have an impact on all components. For example, snowfall could limit the detection range of the sensors. Therefore, a surround-sensor based localization system may require the car to be driven more slowly due to the changed appearance of the surroundings and the low-friction road surface.

Main challenges

The examples derived from the specification of highly automated functions show clearly that we still have major challenges to overcome over the next years. We currently identified six focus areas: a highly robust 360° surround sensing for all use cases, a system architecture, security and reliability with regard to technical failure, precise and up-to-date map data, reliable but affordable validation procedures for functional performance and finally, an adequate regulatory environment. On the other hand, we can assume that the fundamental technical challenges will largely be resolved by the end of this decade.

Validation

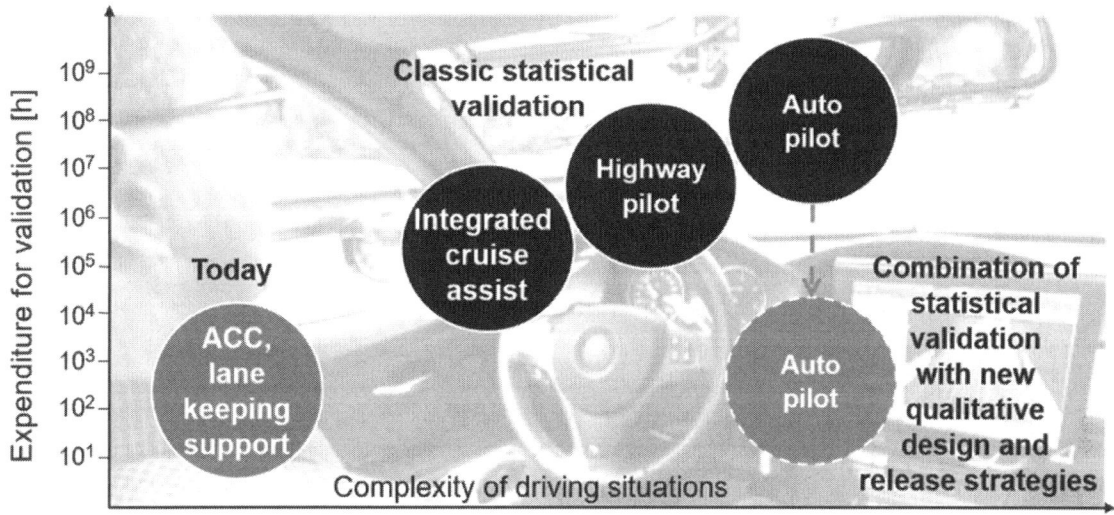

Effort for validation will increase from 0.5 to approx. 550 k man years

Figure 8: Highly automated systems require new release strategies – traditional statistical validation is not suitable for higher degree of automation

The validation of automated driving functions represents a major challenge (see figure 8). In accordance with common methods, a highway pilot needs to complete several million test kilometers before it can be approved for series production. Bosch is working on completely new approaches here.

The effort for a conventional validation would represent approximately 10^2 to 10^4 hours. In the future, partially and fully automated functions would need more than 100,000 years with the same release systematics. As this is completely implausible, we will combine the classic validation methods with new qualitative design and release strategies. This approach is comparable to the procedure used for releasing airliners.

Localization

In addition to robust peripheral detection, highly automated driving requires highly accurate recording of the vehicle position. Localization via GPS is thus not precise enough; it can only provide an orientation. In addition to absolute position fixing, determining the relative position, for example in relation to the lane marking, is crucial. Hence, highly precise and permanently up-to-date maps, as described in figure 9, and alignment with surround sensor data is required. This results in advanced maps with different layers based on conventional maps, which are continuously updated via the cloud.

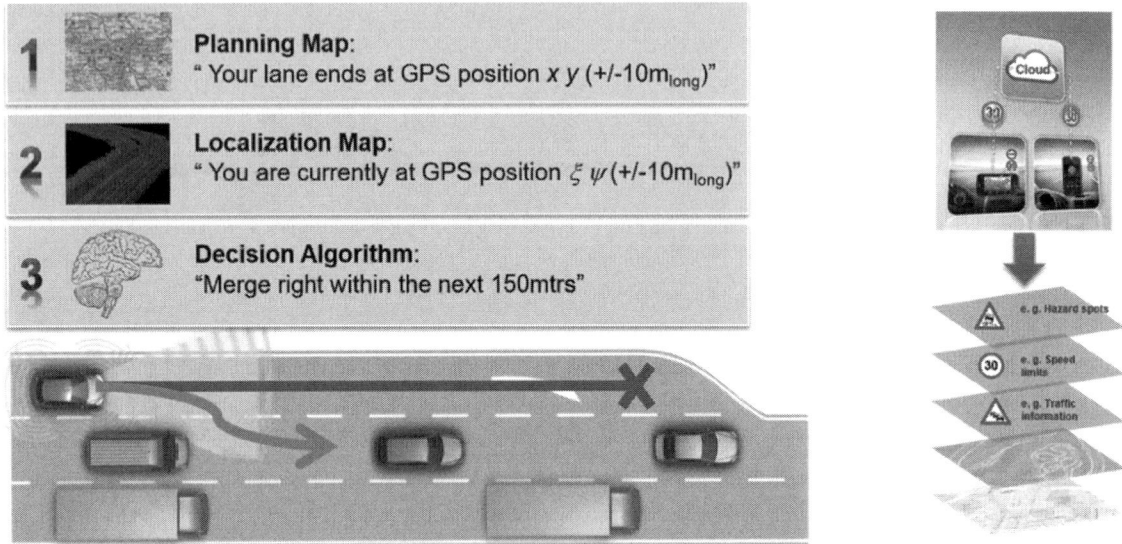

Figure 9: Cloud map data is required – providing long range information for tactical and strategic planning

The different levels are used at different steps during the decision-making process. Localization on road level is required for function activation, e. g. if the function is limited to a certain set of roads such as highways, and in order to determine the route from the current position to the desired destination. Lane level localization is mainly required for lane change use cases, e. g. whether a lane change is possible or whether a lane change is required to reach the navigation goal. Sub-lane level localization is required for maintaining the proper position within the lane. Unavailability of any of these localization levels may lead to degradation of the functional performance or even to a driver takeover request.

Security

The information exchange via the cloud raises the question of data security. Confidence in data security will be crucial in guaranteeing the wide acceptance of networked solutions, including those functionalities required for highly automated driving.

The bowl-model architecture suggested by Bosch and represented in figure 10 comprehends security-protection elements at different levels. This makes it very difficult for a potential aggressor to capitalize on a single weak spot for a serious attack.

Our experience shows that with visible expenditure, a high degree of robustness can be achieved in security. All external interfaces must be run with secure protocols acting like firewall-filters.

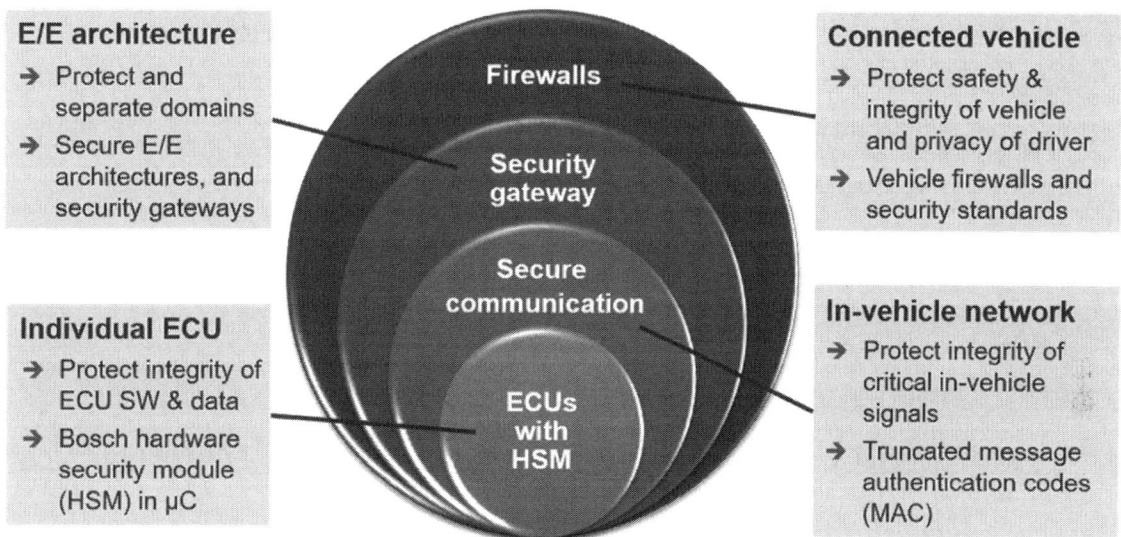

E/E architecture
→ Protect and separate domains
→ Secure E/E architectures, and security gateways

Individual ECU
→ Protect integrity of ECU SW & data
→ Bosch hardware security module (HSM) in µC

Firewalls

Security gateway

Secure communication

ECUs with HSM

Connected vehicle
→ Protect safety & integrity of vehicle and privacy of driver
→ Vehicle firewalls and security standards

In-vehicle network
→ Protect integrity of critical in-vehicle signals
→ Truncated message authentication codes (MAC)

Bosch offers a broad spectrum of solutions for automotive security

Figure 10: Bowl-model architecture

ECU domains, for instance for the drive train, must be separated by precisely configured gateways from the strongly networked multimedia domain. The gateway receives a "firewall function" which prevents aggressors from sending unwanted data packages from one domain to another. The integrity of the communication of safety-data between ECU's within a domain – and also between the relevant sensors – must be ensured in the future. In the case of an aggressor succeeding in infiltrating message packages in the system, they must be recognized and ignored by the system.

Every control device must be secured against individual manipulation, meaning that their software must be authentic at any time and protected against modification by aggressors. Bosch Security modules provide a hardware solution in this sense. They guarantee the protection of the cryptographic keys and procedures.

Sensor set

The surround sensor set for automated highway driving must be capable of reliably detecting all relevant objects in any situation, including during the most challenging use case the vehicle may encounter. The use of redundant sensors with different measuring principles increases the reliability and robustness of the information.

All of them have their limitations for specific use cases. Figure 11 shows some examples that seem quite obvious, taking into account the measurement principle:

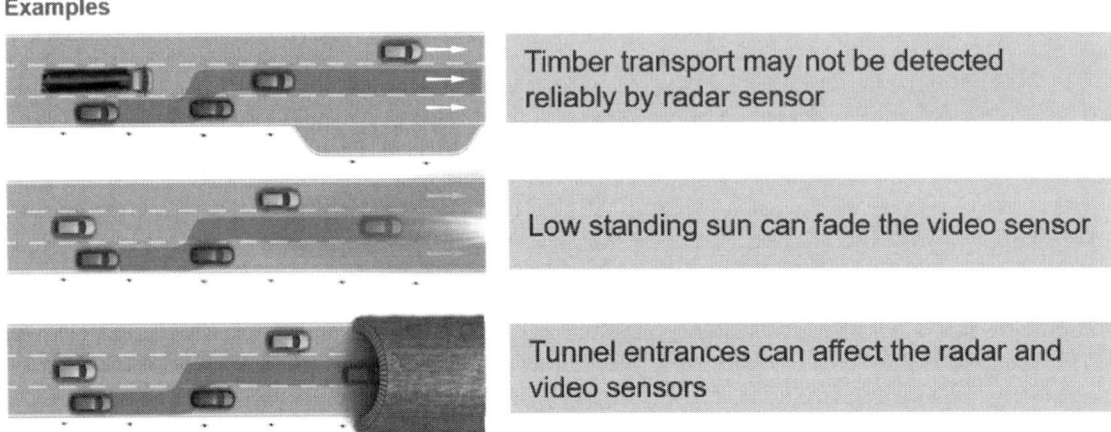

Figure 11: Surround sensing – reliability requirements; highly automated driving raises new challenges for sensor concept

The perception system combines all sensor measurements into a consistent representation of the surrounding world via fusion algorithms. In addition to a representation of all known objects, it also needs to provide a notion of the unknown for many use cases. For example, if the field of view is limited due to fog or obstructions, the system needs to adapt its speed accordingly, because there may be undetected obstacles outside the field of view.

System architecture

Automated driving not only impacts the surround sensing set, but the entire system architecture – the powertrain, brakes and steering. The overall system will change from a fail-safe to a fail-operational architecture. The control units require a significantly higher computing capacity and must be even more closely networked. In the event of a fault, the mandatory basic functionality of all safety-related systems must therefore be guaranteed – at least until the driver has received an appropriate warning and is able to take over the vehicle and respond to a technical problem. This needs to be monitored in order to know at all times whether the driver is able to take control of the vehicle again at all.

Moreover, safety-related vehicle components such as brakes and steering place special demands on the vehicle architecture. These need to be designed in redundancy in order to guarantee absolute reliability, even in the event of a component failure. With iBooster, the electro-mechanical brake booster from Bosch, and the ESP® brake control system, it is already possible today to brake a car independently without the driver having to intervene. Like ESP®, the iBooster can also build up brake pressure completely independently. It is even three times faster than the brake control system and can thus bring the vehicle to a halt sooner in critical situations.

This exemplary set-up is shown in figure 12. During normal operation (both systems are free from errors), the ESP® takes the tasks of vehicle stability control including the processing of vehicle deceleration requests (issued by the automated driving functionality). In case of a fault that forces the ESP® into degraded mode (or even fail safe mode), the automated driving functionality is switched to degraded mode. In this scenario, the iBooster performs the task of bringing the vehicle to the safe state. This concept is called dynamic hot standby.

The mechanical push-through that is available in today's braking systems becomes obsolete in the automated driving mode, as there is no longer a driver available to take over control. The same is true for the steering system; therefore, a fail-operational electrical steering is also required. This can be realized by combining electronic power steering with ESP®, which can issue a yaw momentum through braking of individual wheels.

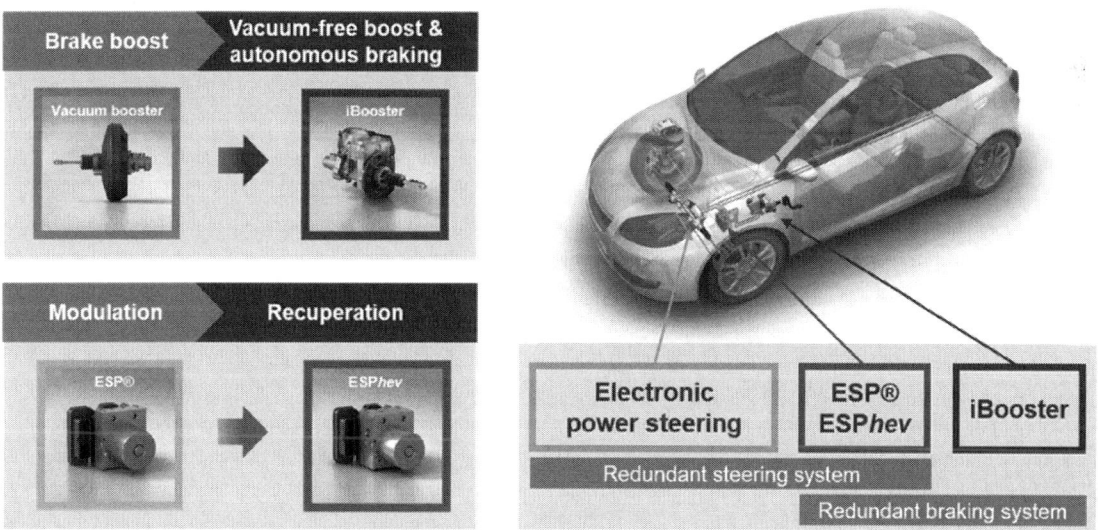

→ Modular actuation concept offers a perfect solution for automated driving

Figure 12: Safety – reliable actuation elements; redundant steering, braking, and stabilization systems required

Regulation

In the future, there are still a few legal hurdles that also need to be overcome. The Vienna Convention of 1968 forms the basis for traffic law in Central and Eastern Europe, Brazil and several other countries. Article 8 of the treaty states that: any moving vehicle must have a responsible driver, that the driver must constantly be in control of the vehicle, and that the driver must refrain from all other activities while driving. Article 13 of the convention states that: every driver of a vehicle must, in all circumstances, have the vehicle under control to be able to exercise due and proper care and to be, at all times, in a position to perform all required maneuvers. The introduction of automated driving thus necessitates a change in the legal framework.

The process of amendment is fortunately already in progress: after the adaptation of the Vienna convention on Road Traffic dated October 2015, which came into effect in March 2016, automated driving functions are permitted as long as the driver can actively override them or turn them off. However, what the driver can do when the car assumes the task of driving is still under debate. Clarification of the regulatory laws is still necessary. In addition, in the area of approval law, an informal working group from the UNECE (the United Nations Economic Commission for Europe) is addressing the R79 regulation, which currently only allows automatic steering intervention at speeds of up to 10 kilometers per hour. The first results are expected by the middle of 2017.

The regulatory framework in the USA is different from the European one. The National Highway Traffic Safety Administration (NHTSA) makes provisions via Federal Motor Vehicle Safety Standards. NHTSA announced in February 2016 that it will consider modifying the definition of the driver in response to changing circumstances: if no human occupant can drive the vehicle, it is more reasonable to identify the driver as whatever instead of whoever is actually driving. This is an important change of perspective that might significantly impact further rulemaking in the United States. Complementarily to federal regulations, several US states have permitted the use of self-driving vehicles in order to promote their development.

Conclusion

In this lecture, we provided a high-level overview of our development of highly automated driving systems. We illustrated challenging situations and use cases and outlined their impact on system design, key technologies and their technical realization.

Automated driving is gradually arriving; it will increase road safety, synchronize traffic flows better, and thus improve efficiency. It will irreversibly change vehicle architecture and will require a new regulatory framework.

We are convinced that automated driving is becoming a reality because it is "invented for life".

How much automation do we really need?

Dr. rer. nat. Karl-Heinz Glander, Lex van Rooij,
ZF TRW – TRW Automotive GmbH

© Springer Fachmedien Wiesbaden GmbH, ein Teil von Springer Nature 2018
R. Isermann (Hrsg.), *Fahrerassistenzsysteme 2016*, Proceedings,
https://doi.org/10.1007/978-3-658-21444-9_2

Automated driving for today's drivers

ADAS functions today and near term

Nowadays many OEMs equip their cars with Advanced Driver Assistance System (ADAS) functions (see figure 1) which assist the driver e.g. to stay in the lane, keep an appropriate distance to the vehicle in front or identify if someone is driving in the driver's blind spot. These functions have a mere informative character or only affect a single task in the driving process within comfort boundaries. Moreover, the driver is always able to overrule them.

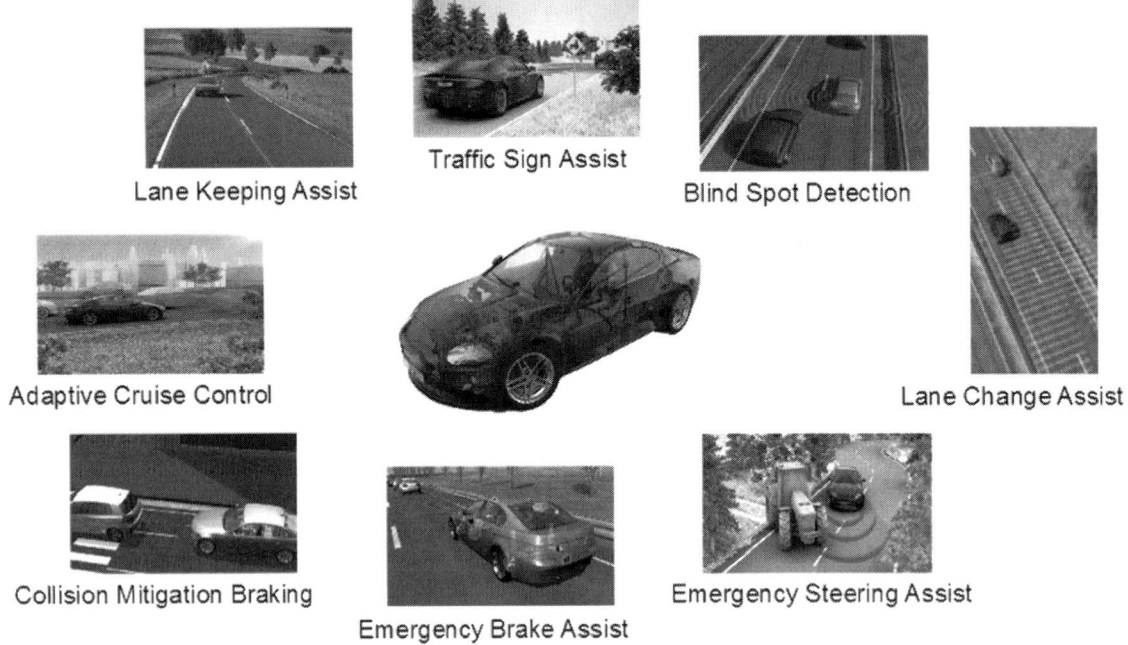

Figure 1: ADAS functions, SAE/VDA Automation Level 0 and 1

Apart from that, so-called emergency assists help the driver in dangerous situations which can be harmful to the driver or to others. They are designed to avoid or mitigate an impending collision by hard braking or by steering manoeuvres. Once initiated these functions cannot be stopped by the driver.

All these ADAS functions are well established and accepted in the consumer market and can be realized by single or dual sensor solutions. As we evolve towards higher automation levels as defined by SAE[1] or VDA[2] (see figure 2), the driver gets decreasingly less

---

1   SAE INTERNATIONAL STANDARD J3016

involved in the driving process. His role gradually changes from a fully responsible, not assisted person towards a mere passenger at full automation level. The necessary framework to realize increased levels of automation will consist of multi-sensor solutions to provide fail-operational driving for at least a certain time period and in specific driving scenarios. The last step currently identified is where a vehicle realizes the complete trip fully automatically, as such there might be no manual controls installed at all.

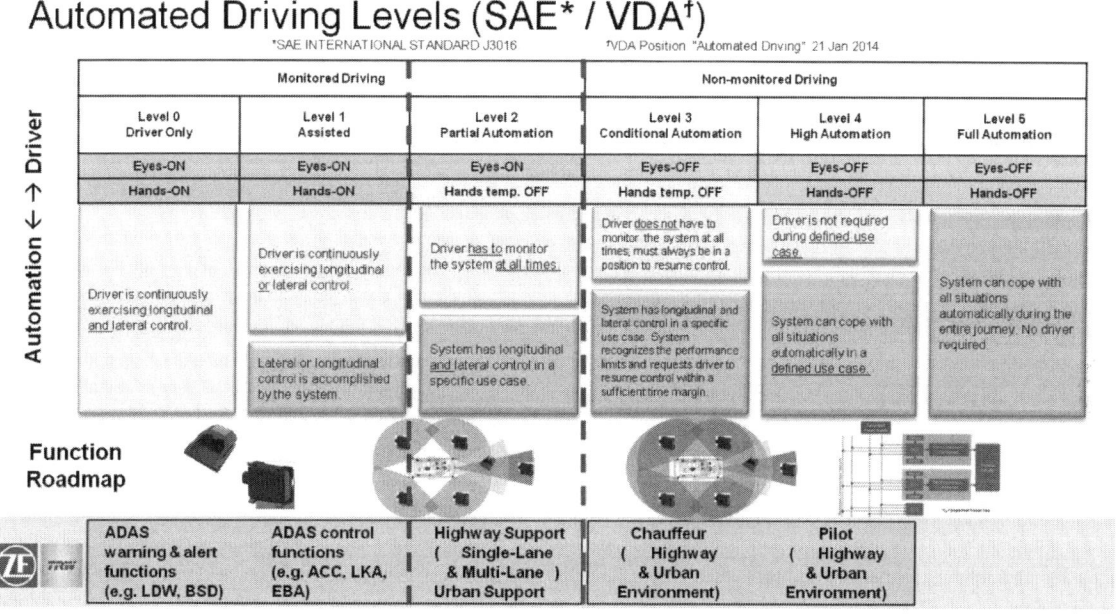

Automated Driving Levels (SAE* / VDA†)

*SAE INTERNATIONAL STANDARD J3016   †VDA Position "Automated Driving" 21 Jan 2014

	Monitored Driving			Non-monitored Driving		
	Level 0 Driver Only	Level 1 Assisted	Level 2 Partial Automation	Level 3 Conditional Automation	Level 4 High Automation	Level 5 Full Automation
	Eyes-ON	Eyes-ON	Eyes-ON	Eyes-OFF	Eyes-OFF	Eyes-OFF
	Hands-ON	Hands-ON	Hands temp. OFF	Hands temp. OFF	Hands-OFF	Hands-OFF

Figure 2: Definition of automation level according to SAE/VDA with ZF TRW's function roadmap

Today conditional automation (level 3) or greater is legally prohibited because the driver is not involved in the driving task, neither in piloting the car nor in monitoring the surroundings. Strictly speaking, there are also legal restrictions for partial automation (level 2) due to the Vienna Convention and the ECE R79. The ECE regulation 79 permits automated steering without hands on the steering wheel to a maximum speed of 12 kph – sufficient for semi-automated parking systems today but not even for traffic jam assist systems.

---

2   VDA Position "Automated Driving" 21 Jan 2014

How much automation does the driver need and how much does he want? Evolution or revolution?

In recent years numerous automated prototype vehicles have filled the media, equipped with the necessary high-tech equipment, which helped to stimulate the current discussion on automated driving. This brings up the question how to proceed with the development of automated vehicles: either adopt the effort to revolutionise the automotive market with vehicles capable of full automation ignoring the high expenses, not yet solved legal issues and cyber security aspects of V2X-communication, or evolve step by step concentrating on the driver's will to accept new features which lead to a new way of driving (see figure 3)?

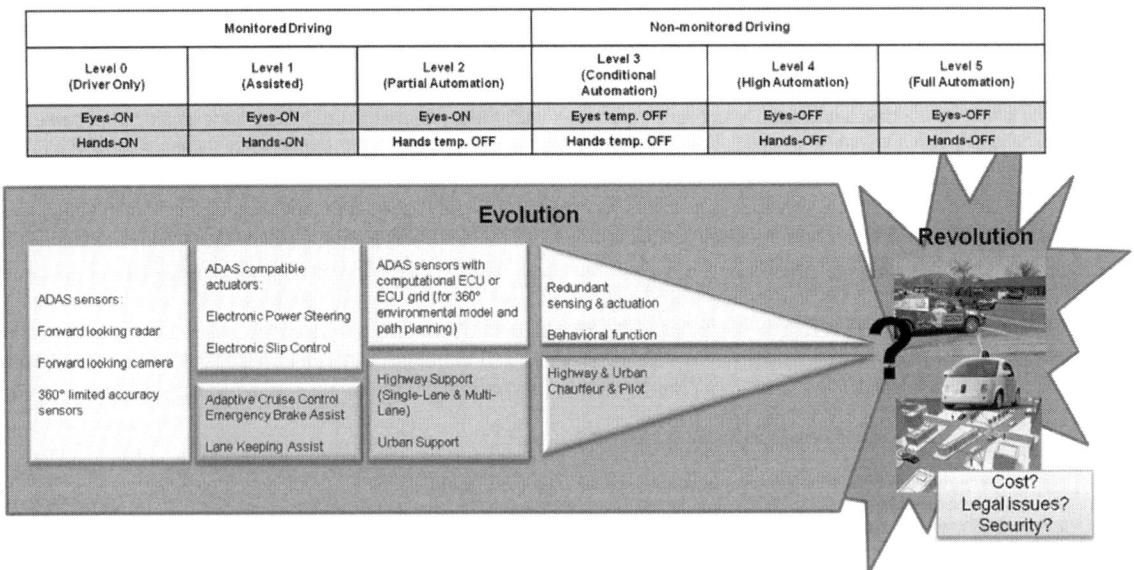

Figure 3: Evolution or revolution: approaches for automated driving

If we chose the latter case, this leads in a first step to the need of ADAS compatible sensors and actuators such as Electronic Power Steering and Electronic Stability Control to realise functions like ACC, Emergency Braking or Lane Keeping. An increasing automation level demands novel vehicle architecture with 360° environmental modelling and path planning with computational ECUs or even an ECU grid where each member ECU takes over specific tasks. This setup can support the driver in highway or urban surroundings providing partial automated driving functionality extending that of current ADAS, which likely results in a high consumer acceptance, also since the driver still is fully responsible during the whole driving process.

The ZF TRW Highway Driving Assist (HDA) function consists of current series equipment like camera, front and corner radar, EPS and ESC combined with PCs capable of embedded software. To expand this to a level 2 Highway Driving Support (HDS) function, the area behind the vehicle has to be covered by two additional short-range radar sensors. The HDS function then provides the driver with multi-lane functionality on highways and with automatic curve negotiation, even on rural roads. The architecture of this function is such that features that improve driver experience can easily be added.

Crossing the border towards conditional automation, for the first time the driver will not need to monitor the environment any more leaving the responsibility of environmental perception to the car alone, at least for a certain time and in specific use-cases. Redundant sensing and actuation are absolutely indispensable to guarantee safe driving from a technical point of view. In addition, there should be a Human Machine Interface (HMI) concept to handle communication and task-handover between driver and vehicle. For complex traffic situations, a behavioural function is required that determines the driving strategy based on the situational analysis.

The driver in the centre of development

In automated driving the vehicle takes over more and more tasks usually performed by the driver, making the key difference to pure ADAS. Therefore at some point, the vehicle becomes the driver, and it is anticipated to perform to a much better degree than a human ever could.

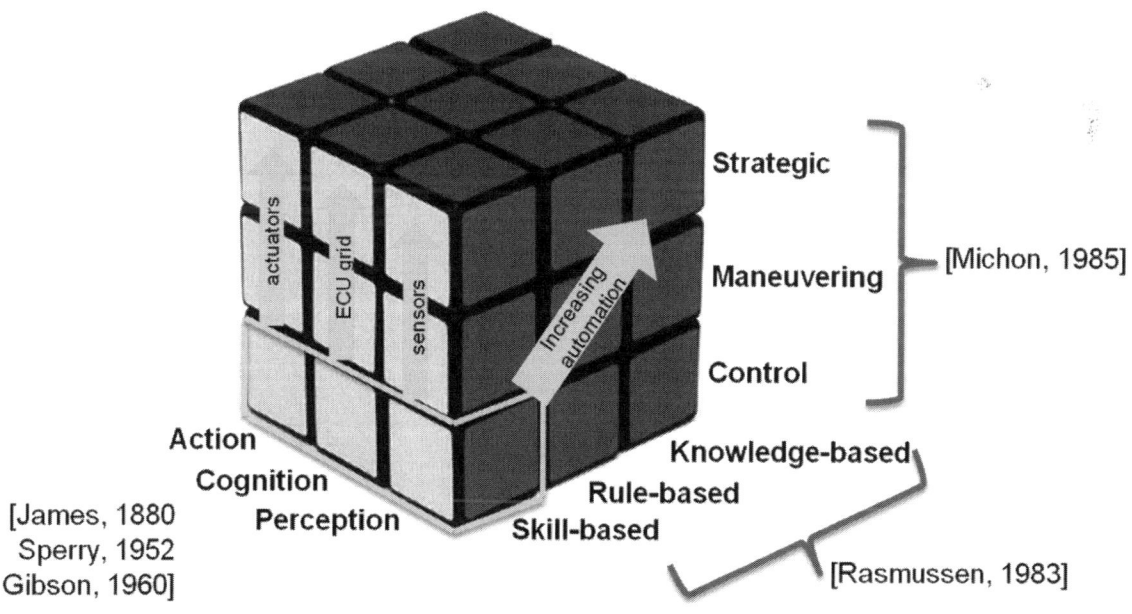

Figure 4: 3D model of driving due to behavioural psychology

Planning to emulate a complex system like a human being driving a car, "human driving" can be segmented in three dimensions, as shown in figure 4.

The first dimension is the physical level consisting of perception with human sensors like eyes or ears, cognition by processing these information in the brain and taking action based on the gathered evaluations. Transferring this to a non-human driver means that the physical level has to be covered by sensors for perception, ECUs for cognition and actuators for action.

The second dimension concentrates on the driver's performance. According to J. Rasmussen[3] there are three levels of skilled performance judging by the nature of behaviour required to fulfil a task and the complexity of processing the related information in the brain. In its simplest form, a skill-based behaviour is needed e.g. to stay in the lane, i.e. performing repetitive tasks without great attention. More complex situations like applying traffic rules can be handled by rule-based behaviour processing sign-like information. The last and most complex behaviour is knowledge-based and should be used for strategic decisions like route planning.

Finally the task of driving itself can be divided into a mere control level with automatic actions, a manoeuvring level with a controlled sequence of actions and a strategic level for planning actions.[4]

Looking at a level 2 system the main objectives are to optimally engage the driver in the highest level driving task while relieving him from repetitive skill-based control tasks such as lane-keeping and car-following. Since the driver feels ultimately responsible for the vehicle's performance he has confidence that the system performs the automated tasks under the condition that he is well-informed about the system's status and mode and the important decisions it takes.

The "Human-Robot-Interaction"

Mimicking the "human way of driving" makes the vehicle seem to be more like some sort of "robot item" because now it has to have a suitably large "brain" to make decisions on its own. Going from no to high automation, the intensity of driver-vehicle-interaction peaks at the transition from partial to conditional automation (see figure 5) because there has to be a large exchange of information between driver and vehicle.

---

3 J. Rasmussen: Skills, rules, and knowledge; signals, signs, and symbols, and other distinctions in human performance models (1983)

4 John A. Michon: A CRITICAL VIEW OF DRIVER BEHAVIOR MODELS: WHAT DO WE KNOW, WHAT SHOULD WE DO? (1985)

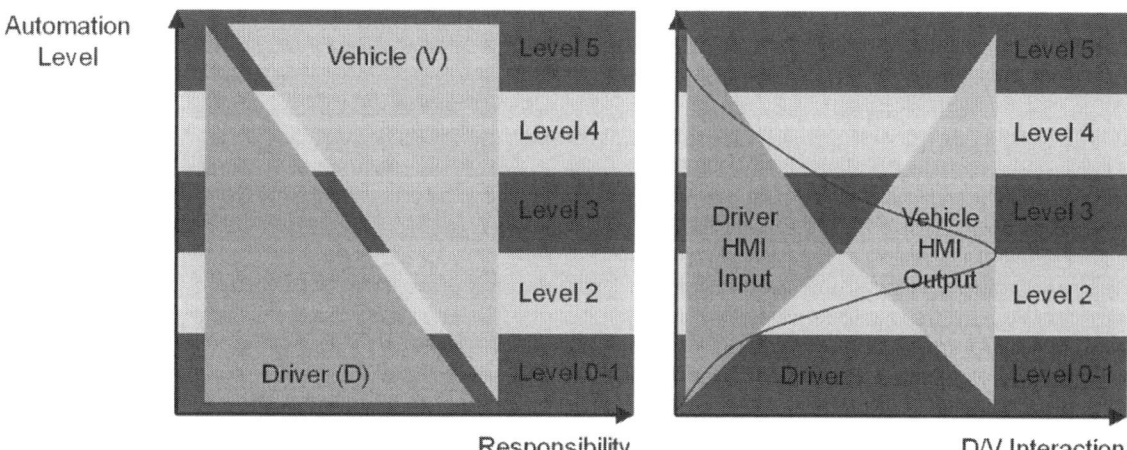

Figure 5: HMI and driver-vehicle-interaction with regard to Automation Level

Assuming the driving robot (hereinafter called the item) is able to detect the driving environment with a 360° field of view and to react on every highway driving situation to a speed up to 130 km/h with the driver as immediate (< 1,5s) fallback option, it can detect objects and scenery as well as plan its driving trajectory. When its status is communicated to the driver via an HMI, this enables him to understand the items' decisions, anticipate on the items' weaknesses and prepare earlier to take over in case of malfunction. Doing this, a level 2 system must not move away the driver's attention from the road. On the contrary, it must be easy to comprehend in addition to incoming information from the current traffic situation, especially once a warning for immediate take over is provided. In the case of a conditional automation level 3 system there has to be a handover procedure within an appropriate time window, during which the system is in charge of the driving task until the driver has fully regained control.

On the other hand, the knowledge of the driver´s attention level, the driver's confidence level to solve the incurring tasks as well as his preferred driving style enables the robot to ensure the driver is monitoring both system and traffic, to pre-deploy its fallback options and to adapt its driving style accordingly. When the driver is fully aware of the traffic situation and the system status, his readiness to act when required will be guaranteed. However, driver attention monitoring is a very sensitive task with regards to privacy and shall not create a feeling of being supervised and nurtured (see figure 6).

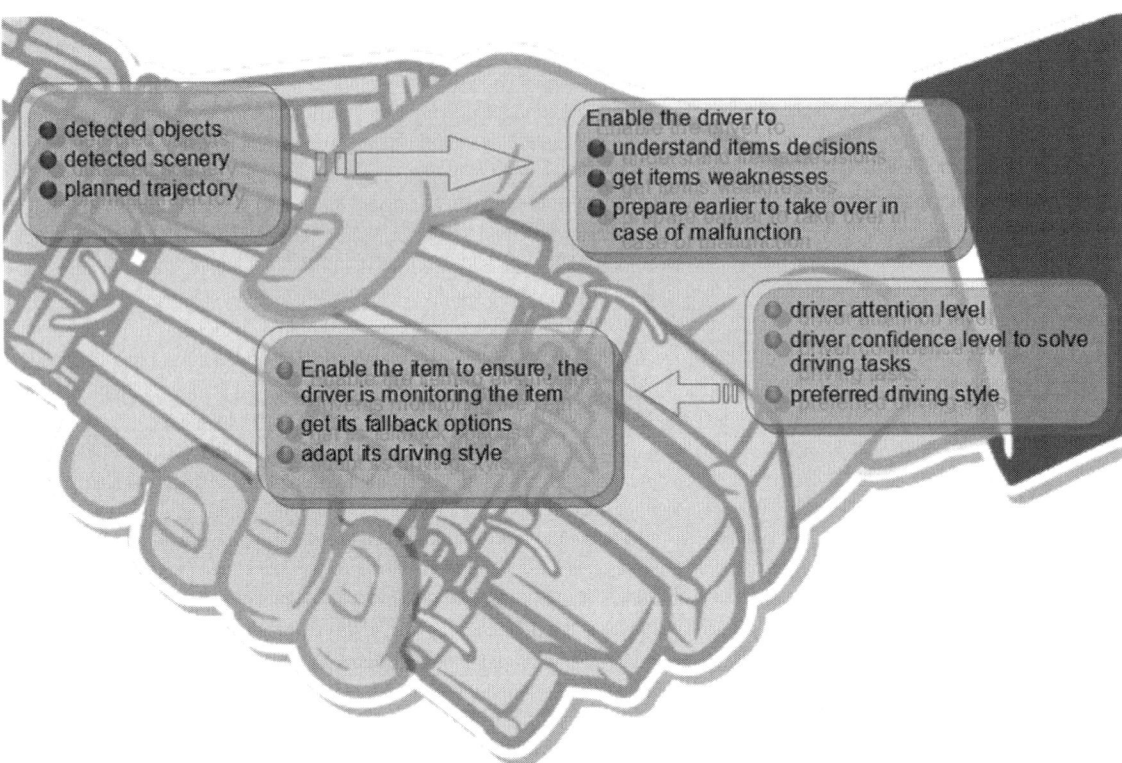

Figure 6: Handshake between robot item and driver forms a new driver-vehicle-interaction

When is the driver performing optimally?

Partial and conditional automation can be a way to less stressful driving, but the crucial point is to keep the driver attentive. On the other hand, there must be a strategy to get the driver back in the driving loop when required without causing an amount of stress he cannot deal with. According to Yerkes-Dodson (see figure 7) the driver's vigilance is perfect at medium arousal: Too much arousal can cause stress and insufficient arousal can be tiring. So, a level 2 or 3 system has to find a healthy balance between inattention and stress.

A fourfold strategy (see figure 7) for arousal to ensure optimal driver performance should contain the following aspects:

1. To relieve the driver by reducing workload and provide physical relaxation.
2. There must be a warning when action is required by the driver.
3. The strategy has to support the driver by providing an energizing task load.
4. At last the driver has to be informed to give confidence in the system, i.e. to avoid stress.

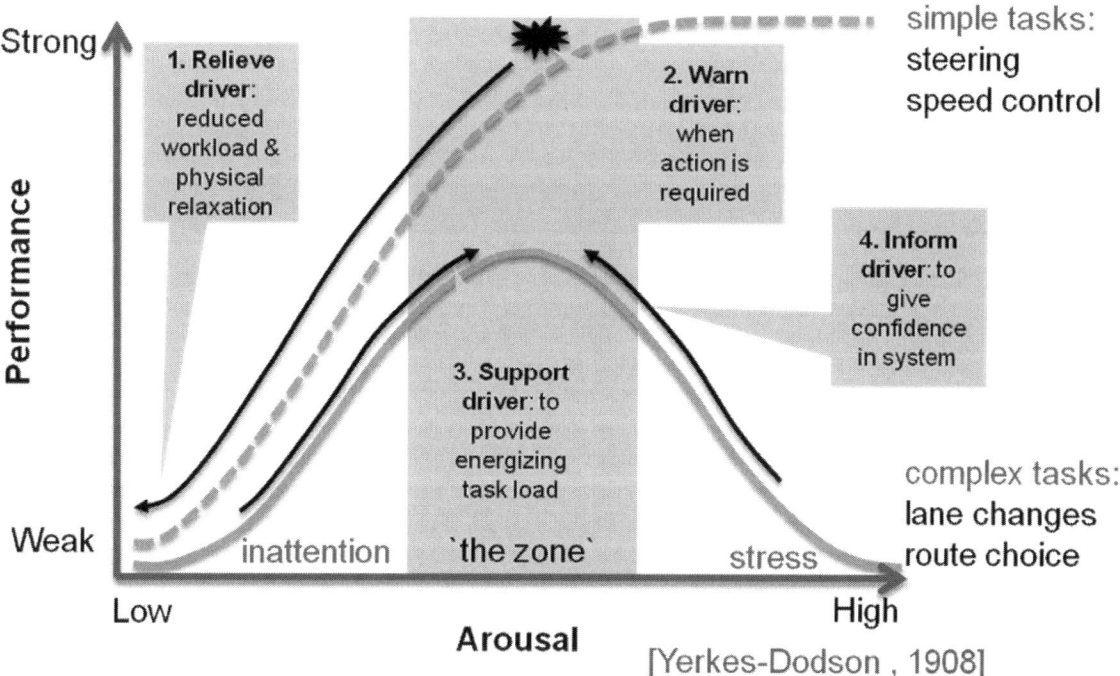

Figure 7: Task performance – arousal relationship.

Driver monitoring

Some ADAS functions like Lane Keeping Assist require hands-on detection in the steering wheel to prevent misuse by the driver. A level 2 system only shows to full advantage in hands-off mode giving the driver the possibility to maximize its stress-reducing effect. During hands-off mode the driver vigilance has to be ensured by some sort of monitoring system. Vehicle sensor based measures which monitor the driving performance, such as car-following performance or steering-wheel based drowsiness detection, are not applicable in automated driving. Physiological measures control heart beat and/or respiration may give an indication on the driver's attention level, however quite indirectly. They cannot give information on whether the driver's eyes are on the road. To the author's opinion, camera-based eyes-on detection offers the most reliable information about the driver's vigilance and his view to the road.

The need of flexible architectures

To enable automated driving, a functional architecture was defined that reflects the processing steps humans undertake when driving (see figure 8). Sensors on the vehicle detect the environment and ego motion. In the perception layer there will be environmental modelling and ego localization, resulting in a scene description. The decision layer analyses the situation and interprets it regarding the necessary driving strategy which is put into dynamic path control in the planning layer. The physical layer at last performs the intended driving task via actuator control.

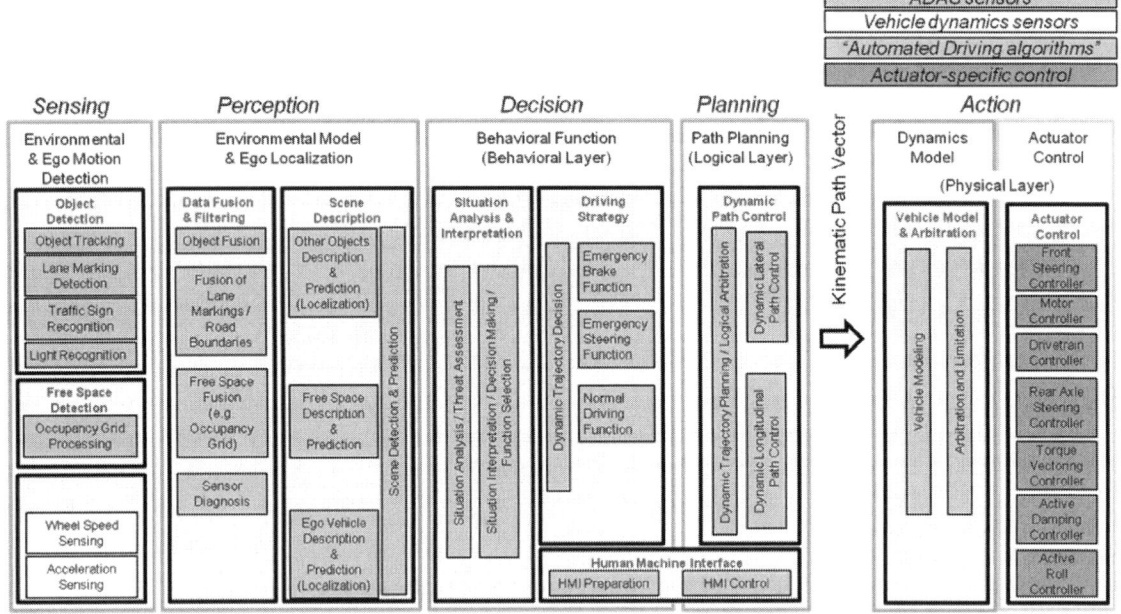

Figure 8: Functional block diagram for automated driving

An automated vehicle will have to detect the environment in 360°. For partial automation, two independent sensors technologies (e.g. radar and camera) might be sufficient where as for a level 3 system triple redundancy is highly recommended (see figure 9). Especially to front direction in minimum three sensors are needed to assure that all relevant objects are safely detected. Whether the redundant sensor technologies require completely different principles (such as camera, radar and lidar) is not yet clearly investigated. It is not unlikely that two different detection technologies within the same principle are sufficient two. Specialized mono-camera plus stereo-camera plus radar or classical automotive radar approaches plus new radar concepts with imaging processing capabilities plus camera can be as sufficient as a set of camera plus radar plus lidar.

10

	Partially Automated	Highly Automated
Forward Direction	• Two independent sensor technologies (e.g. radar + camera) recommended	• Tri-redundancy recommended • Bi-diversification enough?
Rear and Side Direction	• One sensor technology	• Bi-redundancy recommended • Bi-diversification needed?

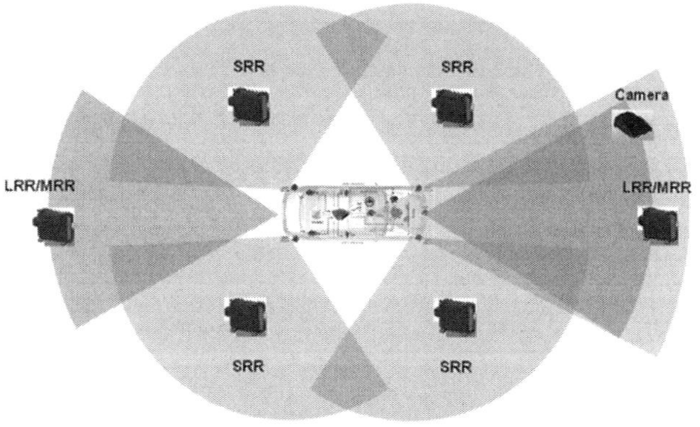

Figure 9: Diversified Redundant Sensing for Automated Driving

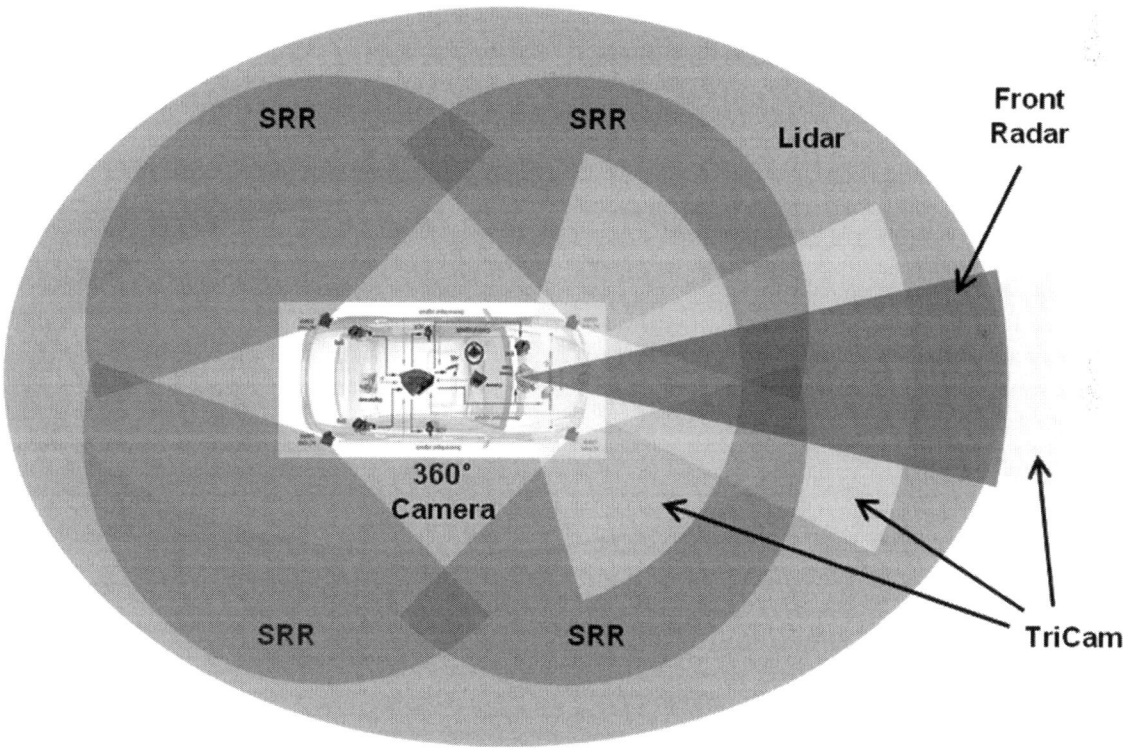

Figure 10: Automated driving three times diversified redundant 360° view

The key strategy for a gradual development of automated driving is a scalable approach regarding the used components. Various blocks in the functional architecture could be distributed over so-called domain ECUs, which are independent of any sensor or mechatronic actuator system. For example, the Image Processing Module could be

11

combined with different ECUs, each of them containing different Automotive Safety Integrity Level (ASIL) functions. The sensor setup for ADAS forward applications realised by front multi-sensor fusion could be easily extended to 360° surround view for partial automated driving by adding backward looking short-range radars (SRR), which may be available for ADAS applications as well. For higher automation levels the sensor setup for a three times diversified redundant 360° view could look like that shown in figure 10. Front radar, a triple lens front camera and lidar monitor the environment ahead. Short range radars on each corner, a 360° camera system as well as a potentially 360° lidar complete the redundant sensor environment.

Summary

This paper shows a potential function and architecture roadmap for deployment of automated driving in the market. The key is not to aim for the last and most difficult automation level now, but to put oneself in the driver's position. From this point of view the next logical step after ADAS is a partially automated driving function which makes use of ADAS-based sensors and actuators in or close to series production. On this level security-sensitive and connected functionality is not required since there is no V2X-communication required yet. Nevertheless, a level 2 driving function can provide the driver with a comfortable hands- and feet-off experience on highways, while the system ensures the driver remains in the loop via involvement in certain tasks, such as lane change initiation or confirmation, as well as via monitoring.

The best choice to realise partial automation is a scalable and flexible component architecture which supports both premium highly-automated driving as well as long-term mass market ADAS and automation by using novel ECUs with specific functions, complying with up to ASIL D and AutoSAR used in a grid in combination with on-sensor processing and redundant actuation.

Influence of Advanced Driver Assistance Systems (ADAS) on damages and repair costs

Helge Kiebach, Gerald-Alexander Beese

KTI GmbH & Co. KG, Lohfelden

© Springer Fachmedien Wiesbaden GmbH, ein Teil von Springer Nature 2018
R. Isermann (Hrsg.), *Fahrerassistenzsysteme 2016*, Proceedings,
https://doi.org/10.1007/978-3-658-21444-9_3

Introduction

Over the last years, ADAS fitment on vehicles highly increased. ADAS functionalities base on surround sensors such as RADAR, LIDAR, Video, infrared, and ultrasonic. But, for various reasons, sensor abilities to sense and interpret the surroundings can get lost about the lifetime of cars. Possible causes are ageing, fault, disassembly respectively assembly of sensors without adjustment, and collision.

Therefore, as part of research activities, KTI has engaged intensively with the topic ADAS and their impact on cost of claims.

Location of ADAS sensors

ADAS sensors are mounted on the vehicle's surface (see Figure 1). That means more expensive components are fitted around the car. Radar sensors and case by case IR cameras are mounted at the front or rear of the vehicle. Thereby, they are vulnerable in case of a crash. The camera and LIDAR sensors are typically mounted behind the windscreen so they are protected in an accident. However, there is a risk for failure malfunction of camera and LIDAR based functionalities as consequence of stonechip.

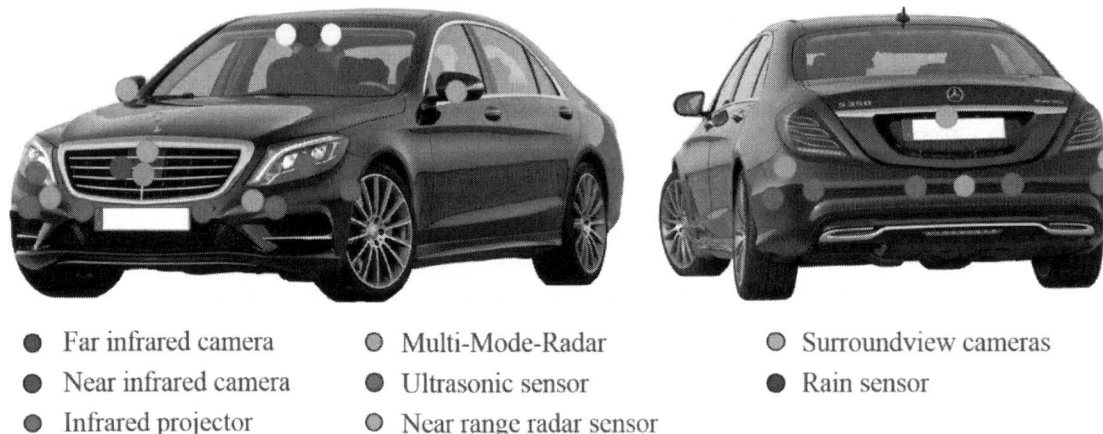

● Far infrared camera	○ Multi-Mode-Radar	○ Surroundview cameras
● Near infrared camera	● Ultrasonic sensor	● Rain sensor
● Infrared projector	○ Near range radar sensor	
○ Mono-/stereo camera	○ Long range radar sensor	

Figure 1: Location of environment-monitoring ADAS sensors in a modern passenger car

According to this, different damage mechanisms between various sensor principles exist. For this reason, KTI performed research on the influence of ADAS on damages and repair costs, on the one hand for sensors attached on front respectively rear and on the other hand for sensors mounted in the area of the inside rear view mirror.

Influence on Claims Costs in case of sensors attached on front respectively rear

An expensive sensor (e.g. radar, infrared camera) mounted in the front grille or behind bumpers could lead to a costly repair. In addition to expensive spare parts alignment of sensors are relevant costs as well.

A long-range infrared camera for night vision systems as spare values about 2,000 € to 2,500 € net. A lower-cost procedure is the possibility to replace the lens cover separately (cost for repair kit about 300 € to 450 € net).

Adaptive Cruise Control (ACC) uses radar sensors fitted at the front of the vehicle. Locations of these sensors are critical regarding claims costs. Damaged radar sensors could result in additional repair costs between 500 € and 3,000 € (Figure 2). A possibility to reduce costs are repair kits for sensor brackets.

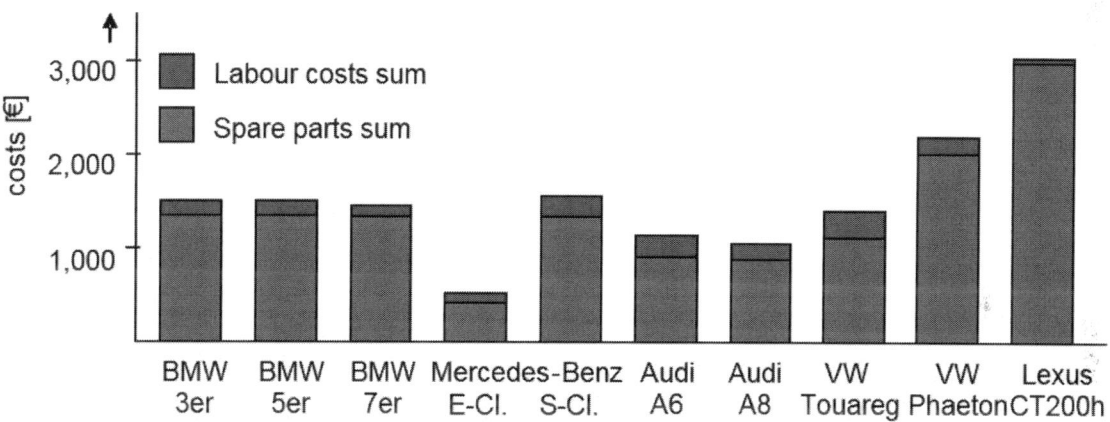

Figure 2: Costs for replacement of front radar sensors at various cars (data status: 06/2013)

Another aspect results in an increase of claims costs: in some cases, OEM service information regarding radar sensors instruct no mechanical damage (e.g. dents) and no body repair work applying filler or painting must be carried out in near about the sensor on the bumper panel. As a consequence, the bumper panel must be replaced even with minor damages in many cases. If a mid-range radar sensor is fixed to the bumper (e.g. sensor for lane change warning), in the majority of cases after replacement or removal of the rear bumper, the system have to be recalibrated.

Influence on Claims Costs in case of Windshield Replacement

Relevance for German Market

According to statistics provided by GDV (Gesamtverband der Deutschen Versicherungswirtschaft e. V.; German Insurcance Association; www.en.gdv.de) for 2011 and 2012 glass claims account for approximately 46 % of claims in Motor Own Damage (MoD), this corresponds to 2.500.000 claims a year. Total claims costs resulting from glass claims amount approximately 1.200.000.000 € every year, corresponding 15 % of all claims costs in MoD. Out of these glass claims, several experts estimate 30 % share of stonechip repairs whereas 70 % windshield replacement. During the last couple of years, various KTI shareholders noticed significant rises in averaged claims costs for glass claims as well as glass claim expenditures exceeding 2.000 € per claim.

Relevant systems and sensors

The study only considers relevant systems and sensors affected in case of a windshield replacement:

– Camera- and video based Systems (KAFAS)

– Laser based Systems (LIDAR)

– Head-Up-Display (HUD)

– Rain-, Lights-, and Moisture sensor (RLS)

Both, radar based and ultrasonic based systems are not being considered. Due to the fact, that OEMs chose different brand names for their systems following ADAS functionalities were considered, some of these functionalities may also be realized in different technical ways (e.g. Laser based AEB for low speed situations whereas high speed AEB can be realized camera based and / or radar based):

– Traffic sign recognition

– Lane departure warning (active / passive)

– AEB – Autonomous Emergency Breaking (Vulnerable Road User recognition)

– (Adaptive light systems)

4

Considered vehicles and models (based on data available 03/2014)

Audi	BMW	Ford	Mercedes	Opel	Skoda	VW
Audi A3	BMW 1er	Ford Fiesta	MB A-Klasse	Opel Adam	Skoda Oktavia	VW UP!
Audi A4	BMW 3er	Ford Focus	MB C-Klasse	Opel Corsa	Skoda Fabia	VW Polo
Audi A6	BMW 5er	Ford Mondeo	MB E-Klasse	Opel Astra	-	VW Golf
Audi A8	BMW 7er	Ford Galaxy	MB S-Klasse	Opel Insignia	-	VW Passat
-	-	-	-	-	-	VW Touran
-	-	-	-	-	-	VW Tiguan

Key findings and implications

- ADAS availability: ADAS are available – at least as an optional feature – for 27 of 28 considered models with growing availability as standard and optional fitting.

- Sensor calibration: After replacing a windshield, system calibration for affected sensors (attached to the windshield) is – except for OPEL models – necessary to fulfil a professional repair following OEM repair guidelines. Remark: this statement above is to change due to vehicle lifecycles, updates, and software updates e.g.

- Investment for body shops: Repair and body shops are required to invest into employee training and qualification, corresponding adjustment devices (calibration targets) and corresponding diagnosis tools. Especially multibrand repair and body shops, and only glazing shops (e.g. Carglass/Belron) are affected by this challenge.

- Claims costs: The study shows a significant rise in claims costs for windshield replacement depending on system configuration and ADAS functionalities. The increasing claims costs are due to both, higher parts prices for windshields and additional required work for sensor calibration. Following illustration shows cost comparison for windshield replacement between RLS-fitted (assumption standard fitted) and ADAS-fitted (KAFAS, LIDAR, and HUD) windshields based on an hourly rate of 85 €.

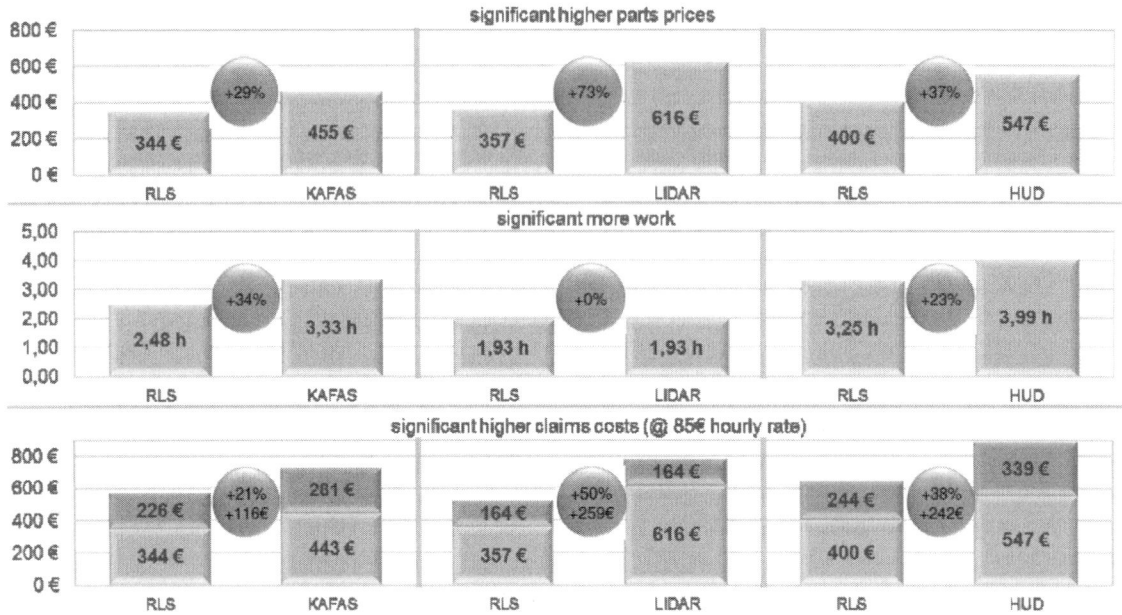

Figure 3: Cost comparison for windshield replacement between RLS-fitted and ADAS-fitted (based on data available 03/2014)

Windshield sensors - Conclusion and Outlook

Current ADAS result in an increase of claims costs for windshield replacement amounting to + 30 % in average, with wide spreads between different OEM and ADAS functionalities. Further development (e.g. Euro NCAP regulations) tend to increase market penetration and fitting rates of ADAS.

On the other hand, ADAS – especially AEB systems – are assumed to have strong positive influences on accident avoidance (declining claims frequency) and accident mitigation (reducing claims severity) resulting in declining claims expenditures in third party liability.

Effects on market participants

Accident research has shown, ADAS help to reduce the frequency and severity of accidents. However, this applies primarily for serious accidents but less for minor accidents. The reason therefor is, that ADAS have been developed to reduce injuries caused by accidents (e.g. Automatic Emergency Braking systems and Electronic Stability Control). But the majority of all accidents, namely accidents without bodily injury, until now are less affected by ADAS. Low speed collisions are the most common sort of accident (see Figure 3). According to this, 80% of all accidents occur with a energy equivalent speed of less than 10 km/h. And in addition a lot of damages on windshields. Apart from that, ADAS in most instances can reduce the collision speed. As a result injury risk will decrease, and vehicles are economical repairable instead of total loss.

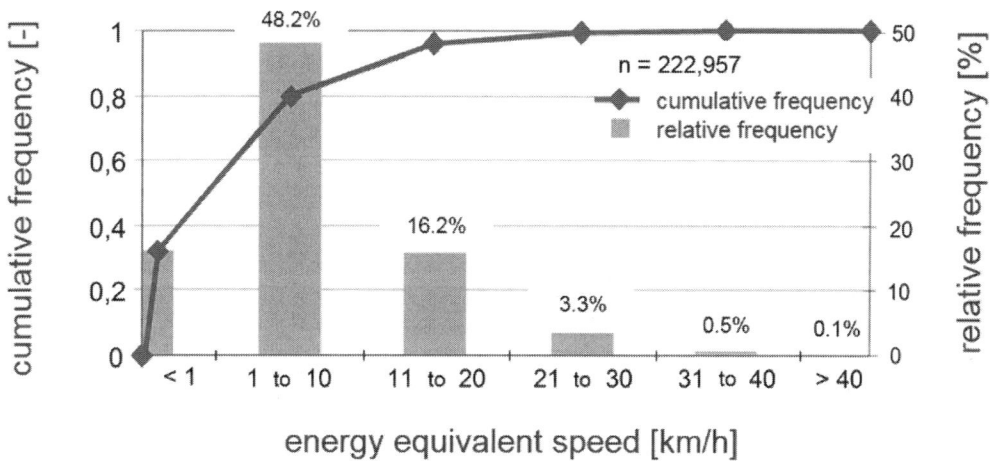

Figure 4: Frequency of impact speed (source: DEKRA accident research group)

Insurers and vehicle manufacturers

For insurers is the point, whether reducing accident frequency and severity or expensive repair costs at minor collisions have more influence on claims costs. In this connection must be pointed out, the impact of ADAS on vehicle claims and premiums will depend on a number of factors (e.g. usage, penetration, ADAS design and performance).

In many countries, insurance rating systems have been introduced. The cost of repairs as a result of low speed crash test criteria (mostly the RCAR standard) plays a major role in deciding the insurance group rating and, therefore, the whole life insurance costs of the vehicle [1]. For example in Germany, motor vehicle insurance premiums (motor liability, semi-comprehensive cover, and fully comprehensive covers) are calculated by separating the individual car types into categories. As of 1996, the classification depends on two criteria; claims frequency as a percentage of total number of claims and average loss across the total vehicle stock. While the claims frequency depends primarily on driver behavior, the average loss can be affected by the vehicle's construction. This system regulates itself because the claims frequency and the average loss of any car type must be a substandard of all cars in the market. At the first classification of a new car type, the claims frequency is as far as possible assumed from the comparable predecessor model. The claim amount is derived directly from the RCAR test, which is then projected on the total of average losses. Since 2010, the RCAR bumper test has served as test procedure by the German Insurance Association in addition to the RCAR low speed test. From an insurance cost view, the requirements are to limit the deformation or damage to structural parts and cost-intensive components such as radar sensor and/or infrared cameras. Summarised, low cost of ADAS sensors have a favorable impact on the running costs of a car. Often, these properties are subject to crash repair tests and are included in the calculation of insurance premiums.

In 2013, GDV has published a paper that describes the consideration of AEB systems at the initial classification procedure. Embodied in the document are, among other issues, the requirements on Automatic Emergency Braking systems (e.g. equipment rate, avoidance of a collision). If the AEB system conform to prerequisites and minimum requirements, the positive effect to be expected will be considered in kind reduction of the initial classification by one type class[1] (motor liability insurance and full comprehensive motor insurance). Consideration of the AEB system will apply throughout the period of validity of the initial classification, and will then be replaced, as hitherto, by statistical data taken from actual claims settlement. When classifying new model series, whose forerunner models had an AEB system that was already considered in the classification procedure, no discount will be granted.

Repair shops

For correct ADAS functions, it is necessary to fulfil a professional repair following OEM repair guidelines, for various brands and models over a vehicles lifetime. This requires a packet of expensive measures, e.g. specific calibration tools per OEM such as appropriate adjustment devices (targets) and diagnostic tools.

Another important precondition for professional repair of ADAS is specific know-how and qualification of the repair shop employee.

---

1   currently the motor liability insurance is class-divided in 16 type classes and 25 type classes exist at full comprehensive motor insurance

Costs for specific calibration tools exemplary selective manufacturers

car manufacturer	adjustment ADAS	required standard tool	required special tool	costs net	action
BMW	ACC	diagnostic tool	ACC adjustment device ACC ground rail track	€ 841.00 € 113.68	radar sensor for ACC adjust
BMW	KAFAS	diagnostic tool			camera adjust
Mercedes-Benz	Distronic Plus („old sensor")	diagnostic tool wheel alignment equipment headlamp beam setter	Distronic Plus adjustment device vacuum pump	€ 294.00	radar sensor Distronic adjust
Mercedes-Benz	Distronic Plus („new sensor")	diagnostic tool			radar sensor Distronic adjust/calibrate
Mercedes-Benz	KAFAS	diagnostic tool headlamp beam setter	adjustment device for night vision system	€ 286.00	Camera adjust
Volkswagen	ACC and KAFAS	diagnostic tool wheel alignment equipment	adjustment device VAS 6430	€ 5,852.00	radar sensor/camera adjust

Summary

During the last years, ADAS fitment on vehicles highly increased. In future, market penetration of ADAS will rise. Accident research shows that ADAS can increase road safety.

Precondition for the possible increase of safety is correct system functioning throughout its entire life. In this context of particular importance is the fact, that the operation of driver assistance systems depends to a great extent on correct sensor information. ADAS use a large number of different sensors. Some are high cost but offer broad functionality and high performance, whereas others are cheaper but have limited functions and restricted performance. Depending on the sensor principle, the quality of this information can be negatively influenced by a wide range of interference factors, such as stonechips, collisions and unprofessional repairs.

Furthermore, sensor replacement cost plays a major role in defining insurance group ratings and therefore the whole insurance costs of the vehicle.

ADAS could significantly reduce accident frequency and repair costs. This has significant effects for market participants regarding accident damage business, such as insurer, car manufacturers and repair shops.

Literature

[1] Leimbach, F.; Kiebach, H.: Reparability and Insurance Ratings in the Development of Cars. In: Encyclopedia of Automotive Engineering, Wiley, 2015

[2] GDV: Consideration of driver assistance systems - Automatic Emergency Braking function – AEB system - at the initial classification procedure (November 2013) http://www.rcar.org/Papers/Procedures/CrashStandards_GermanRatingSystem.pdf

Zukünftige Tests von Fahrerassistenzsystemen im Verbraucherschutz

Von der Unfallforschung bis zum Test

13. April 2016; ATZ-Fachtagung Fahrerassistenzsysteme; Frankfurt am Main

Dipl.-Ing. (FH) A. Rigling; ADAC e.V. Technik Zentrum

© Springer Fachmedien Wiesbaden GmbH, ein Teil von Springer Nature 2018
R. Isermann (Hrsg.), *Fahrerassistenzsysteme 2016*, Proceedings,
https://doi.org/10.1007/978-3-658-21444-9_4

1 Introduction ADAC / Euro NCAP

2 Active Safety in Euro NCAP

3 Future Tests

Introduction ADAC

- Europe's biggest automobile club. 19,2 mio members [02/2016]
- ADAC Technik Zentrum - Landsberg am Lech; since 1997
- Testlab for european automobile clubs
- Development of new consumer protection tests

- Euro NCAP activities
  - Test lab (1 of 7)
  - Member (1 of 12)
  - Accredited for all tests

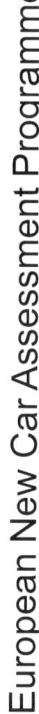

European New Car Assessment Programme

ADAC Test Development

- ADAC Accident Research (n > 19.600)
- AEB Car-2-Car test in 2011 and 2012
- Development of a soft crash target
- Euro NCAP Vehicle Target - EVT
- AEB Pedestrian test in 2013
- Results on www.adac.de

Zukünftige Tests von Fahrerassistenzsystemen im Verbraucherschutz; A. Rigling; ADAC e.V.

4

ADAC

1 Introduction ADAC / Euro NCAP

2 Active Safety in Euro NCAP

3 Future Tests

Zukünftige Tests von Fahrerassistenzsystemen im Verbraucherschutz; A. Rigling; ADAC e.V.

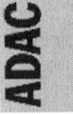

Active Safety in Euro NCAP

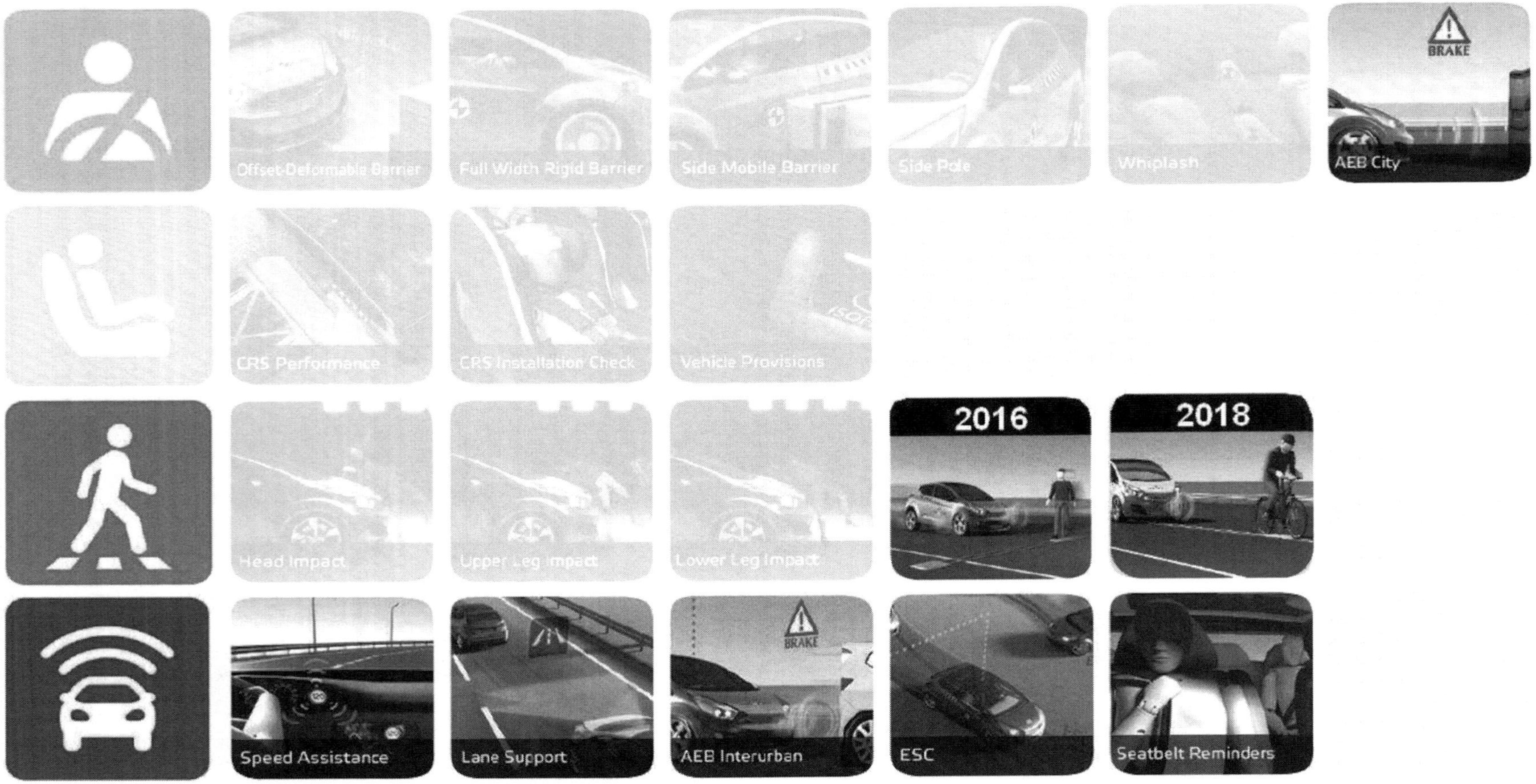

AEB City

Whiplash

Side Pole

Side Mobile Barrier

Vehicle Provisions

Lower Leg Impact

2016

2018

AEB Interurban

ESC

Full Width Rigid Barrier

CRS Installation Check

Upper Leg Impact

Lane Support

Seatbelt Reminders

Offset-Deformable Barrier

CRS Performance

Head Impact

Speed Assistance

AEB Vulnerable Road Users

2015 state-of-the-art system performance:

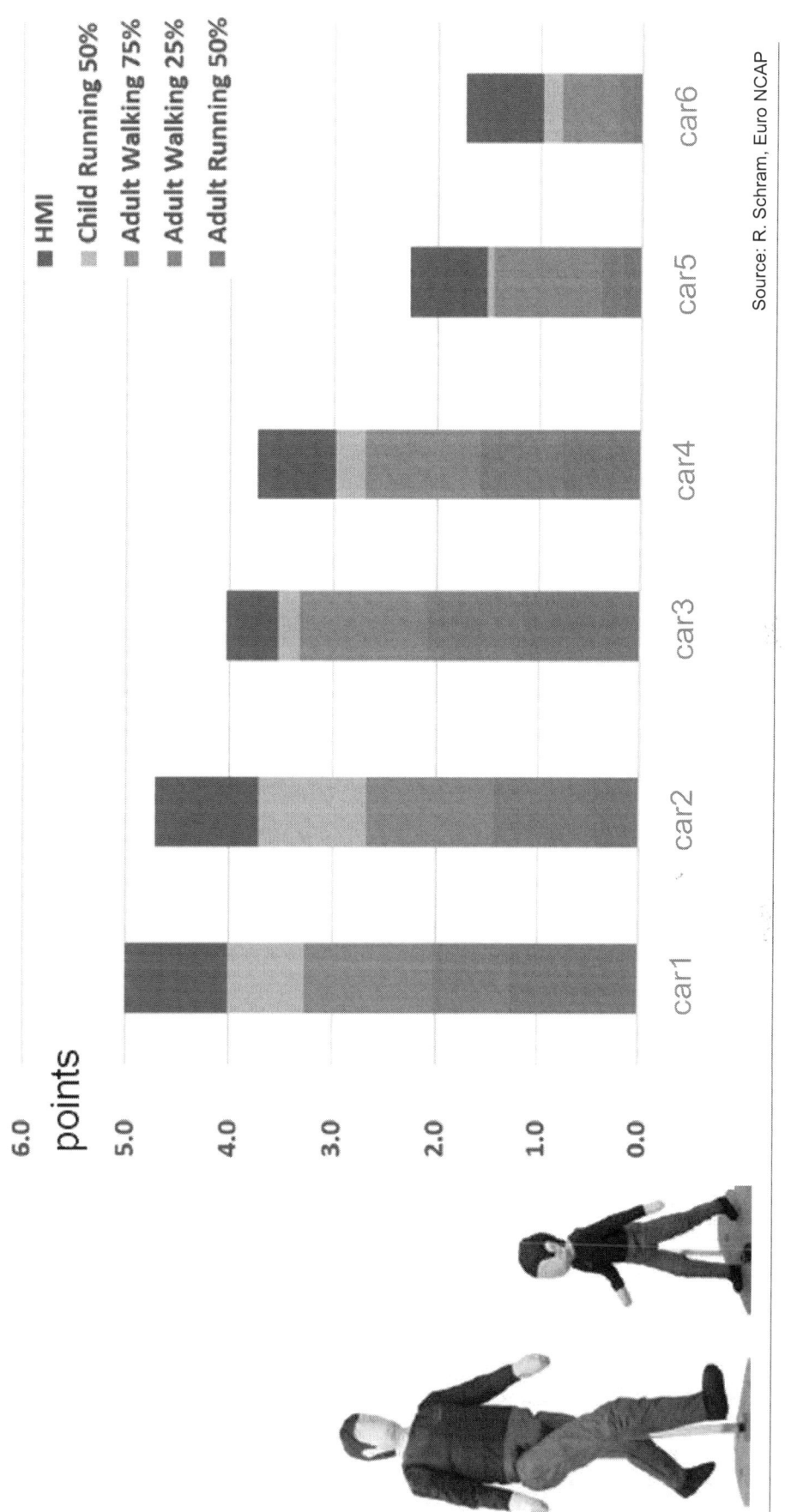

April 2016   Zukünftige Tests von Fahrerassistenzsystemen im Verbraucherschutz; A. Rigling; ADAC e.V.

7

Source: R. Schram, Euro NCAP

Lane Support Systems - 2016

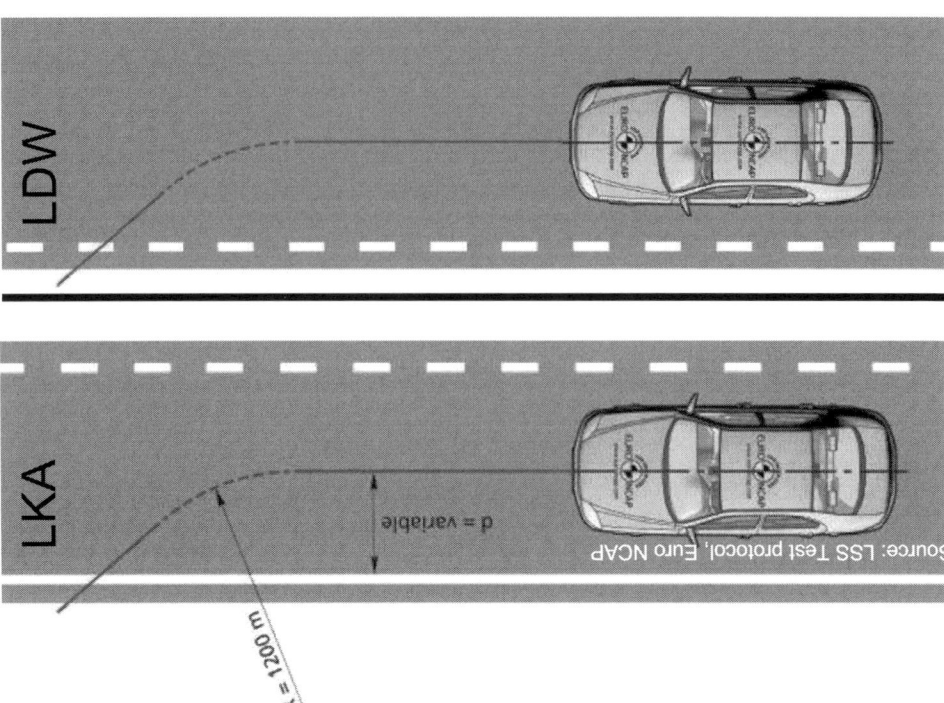

- Defined lateral velocities: 0,1 m/s – 1 m/s

- LDW - Lane Departure Warning

- LKA - Lane Keeping Assist

- LKA system can achieve all points

- Vehicle Speed: 72 km/h

- HMI Points:

  Default on
  Haptic Warning
  Blind Spot Warning

Active Safety in Euro NCAP

- No substitution of passive safety by active systems!

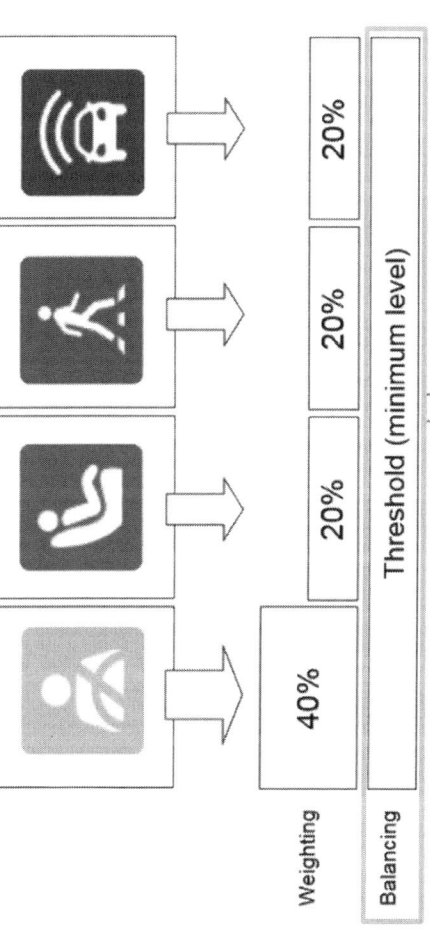

Weighting: 40% 20% 20% 20%

Balancing — Threshold (minimum level)

- 2016 + 2017
  - AEB interurban or LDW/LKD needed for 5 stars
  - Can be achieved by mono camera or mono radar

- 2018 + 2019
  - AEB interurban and Lane Support System needed for 5 stars
  - Can be achieved by high level camera or fusion system

April 2016   Zukünftige Tests von Fahrerassistenzsystemen im Verbraucherschutz; A. Rigling; ADAC e.V.

details see online: www.euroncap.com
Ratings Group Report; March 2015; Version 1.0

9

ADAC

Active Safety in Euro NCAP

- missing AEB interurban fails 5 Stars for small cars in 2015

- Volvo XC 90:
  100% of Safety Assist

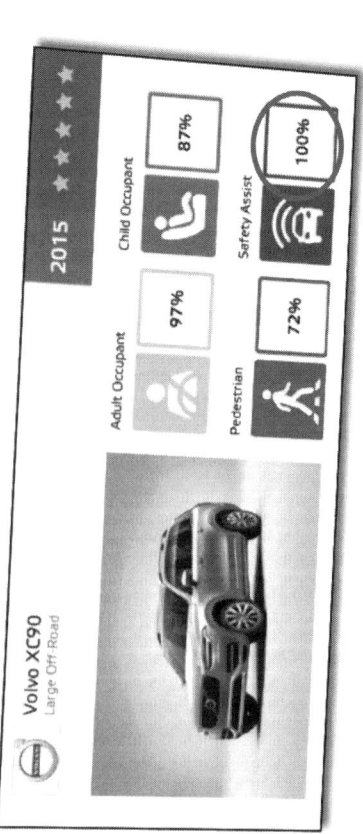

April 2016

Zukünftige Tests von Fahrerassistenzsystemen im Verbraucherschutz; A. Rigling; ADAC e.V.

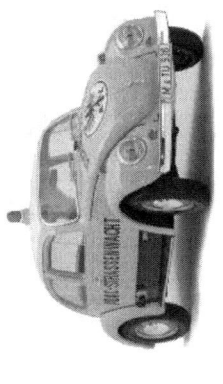

Active Safety in Euro NCAP

From 2016: Dual Rating Option

2015

Base Rating

2016 ★ ★ ★ ★ ★

Standard safety specification

Optional Rating

2016 ★ ★ ★ ★ ★

Standard safety specification +
optional active safety pack

- Base rating min. 3 stars
- Active safety pack available on all variants
- For a limited time only
- Has to be sold on min 25% / 55% of the cars

ADAC

1 Introduction ADAC / Euro NCAP

2 Active Safety in Euro NCAP

3 Future Tests

Zukünftige Tests von Fahrerassistenzsystemen im Verbraucherschutz; A. Rigling; ADAC e.V.

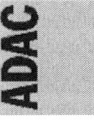

Roadmap 2020

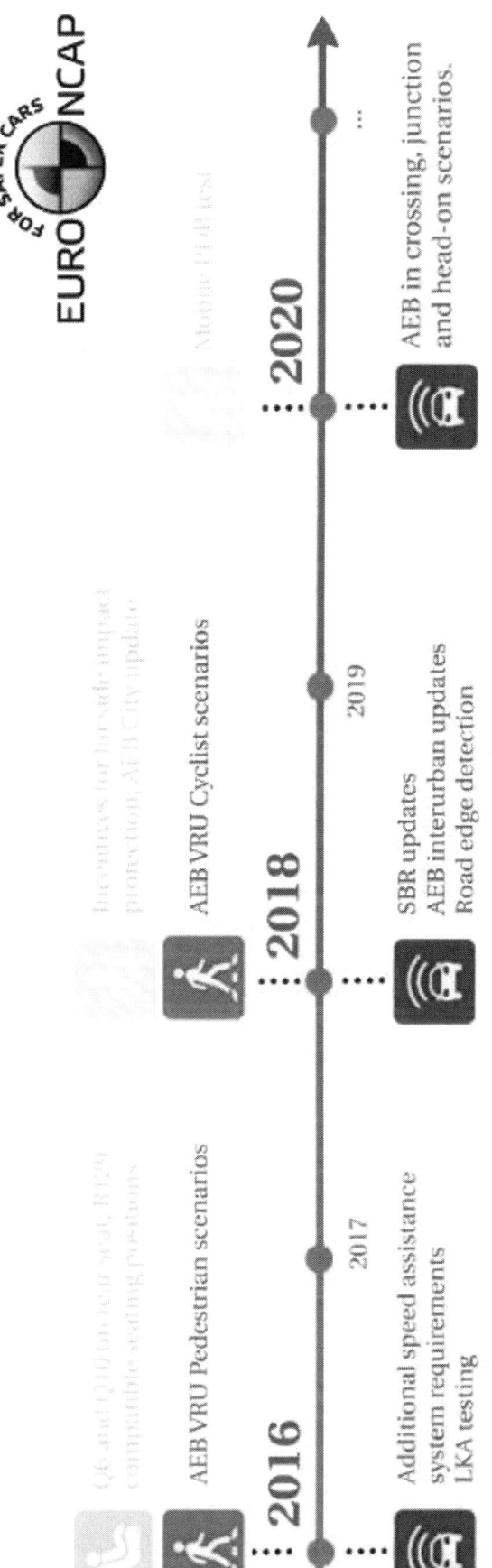

April 2016   Zukünftige Tests von Fahrerassistenzsystemen im Verbraucherschutz; A. Rigling; ADAC e.V.

reference: Euro NCAP

13

AEB Car-2-Car

- AEB Car-2-Car 2018 - 2020
- Cut-In scenario relevant for rear-end scenario has a share of 1.7 % in the actual (until 2014) GIDAS study from BASt*.

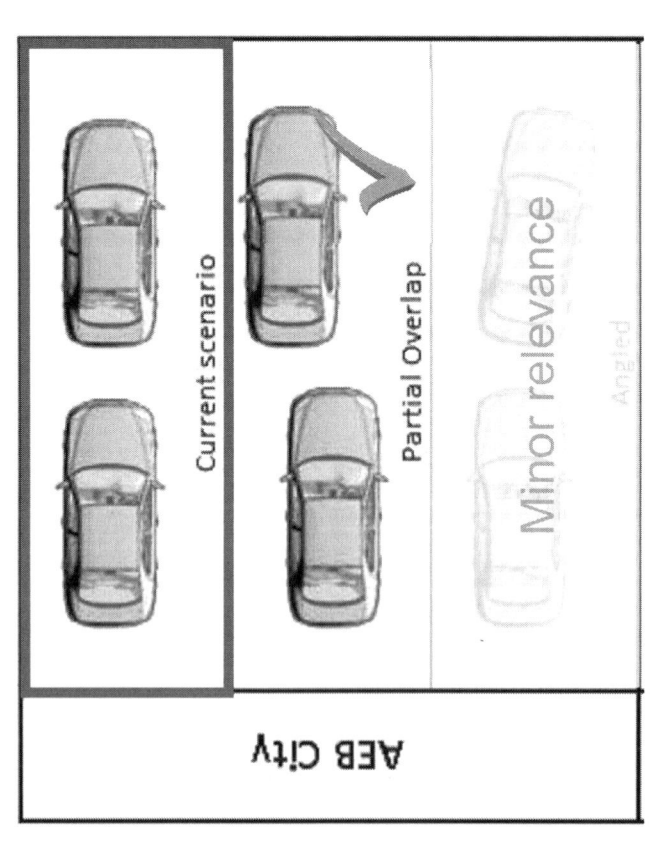

AEB Inter-Urban

Current scenario

Cut-in

2020 — Junction - Crossing

2020 — Junction - Turning into traffic ahead

Lane Support Systems

Head-on

AEB City

Current scenario

Partial Overlap

Minor relevance

Angled

*Quelle: M. Wisch, P. Seiniger, O. Bartels, BASt.

April 2016   Zukünftige Tests von Fahrerassistenzsystemen im Verbraucherschutz; A. Rigling; ADAC e.V.

14

AEB Car-2-Car

- 3-D Soft Crash Target concepts

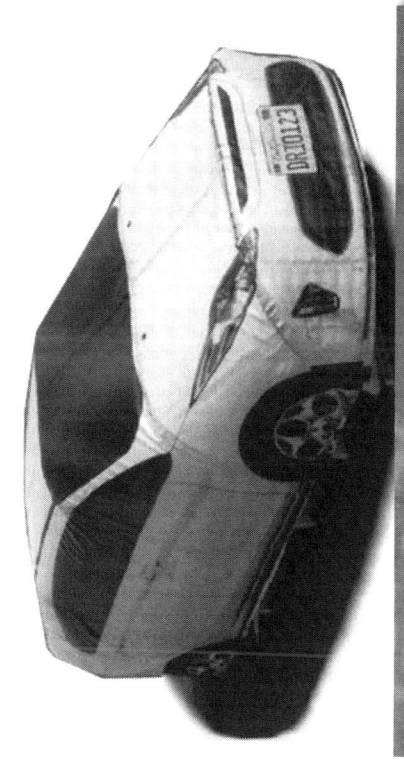

reference: 4A active systems

reference: DRI / ABD

Zukünftige Tests von Fahrerassistenzsystemen im Verbraucherschutz; A. Rigling; ADAC e.V.

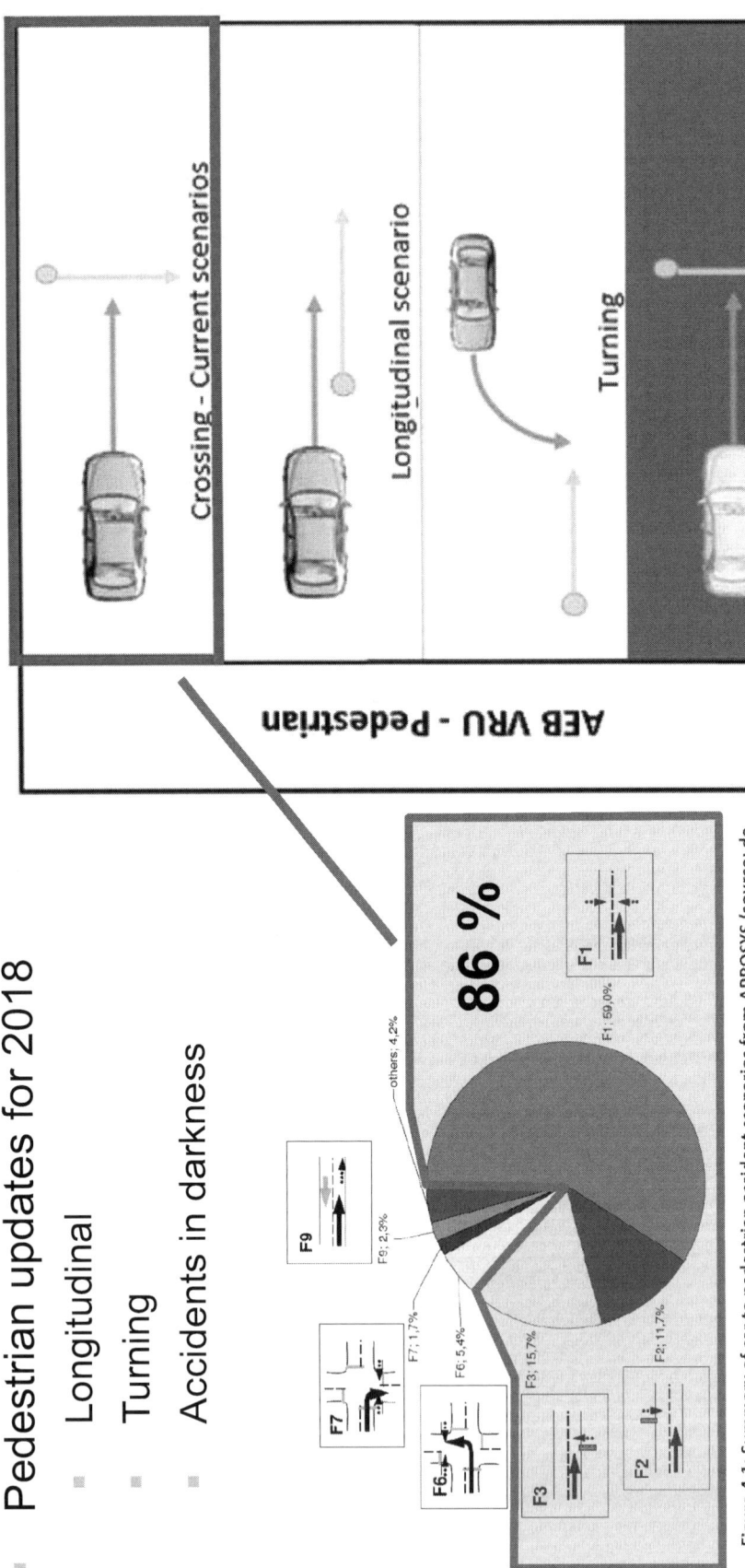

AEB VRU - Pedestrian

- Pedestrian updates for 2018
  - Longitudinal
  - Turning
  - Accidents in darkness

Figure 4-1: Summary of car to pedestrian accident scenarios from APROSYS (source: de Lange, 2007)

Quelle: Scenarios and weighting factors for pre-crash assessment of integrated pedestrian safety systems. M. Wisch, P. Seiniger, C. Pastor, BASt. M. Edwards, C. Visvikis, C. Reeves, TRL

AEB VRU - Pedestrian

- Pedestrian updates for 2018
  - Turning scenario
  - State of the art sensors

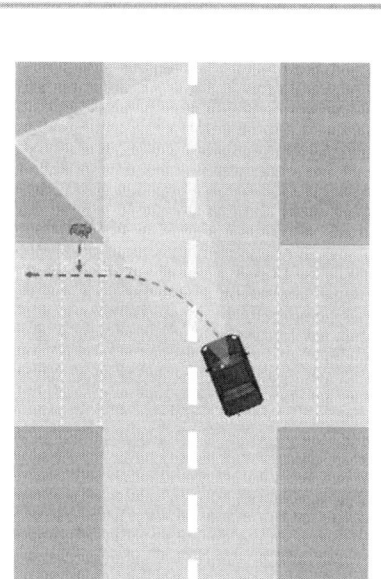

Simulations with IPG Carmaker: P. Wawrziczny; Hochschule Kempten

April 2016   Zukünftige Tests von Fahrerassistenzsystemen im Verbraucherschutz; A. Rigling; ADAC e.V.   17

AEB VRU - Pedestrian

- Options for longitudinal:

  - < 50 km/h – Braking

  - < 80 km/h – Warning

    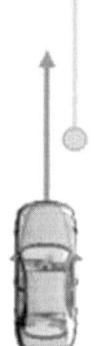

- Accidents in darkness

  - Darkness definition
  - Further test-experience needed

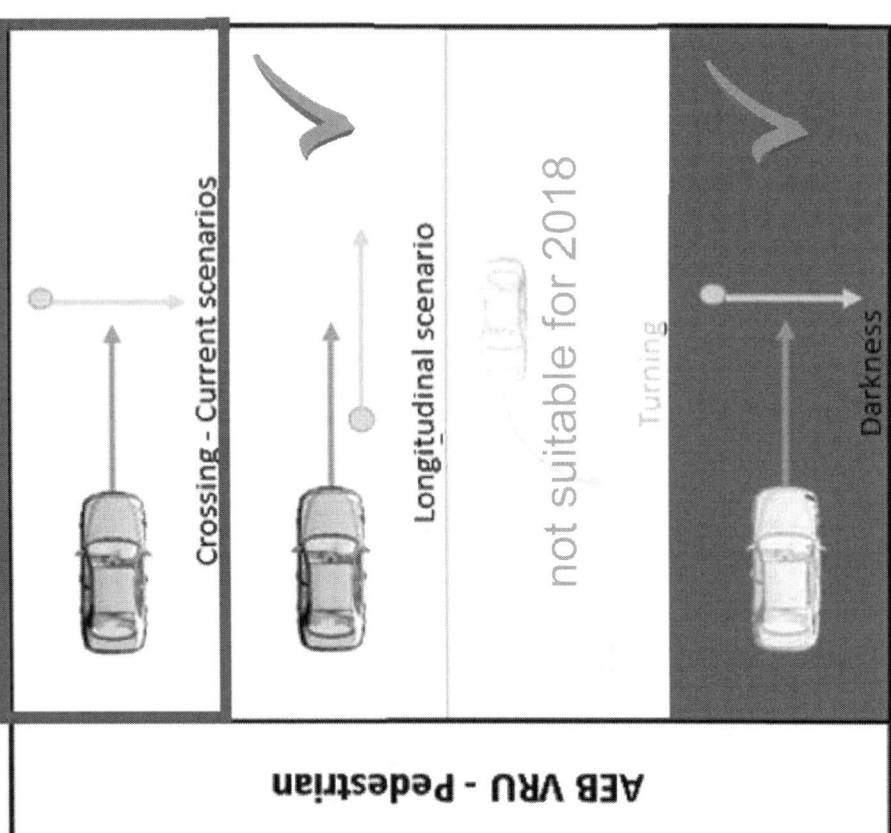

AEB VRU – Pedestrian

Crossing - Current scenarios

Longitudinal scenario

not suitable for 2018

Turning

Darkness

AEB VRU - Cyclists

- Cyclist accident scenarios 2018
- CATS consortium*
  - Crossing
  - Turning
  - Longitudinal

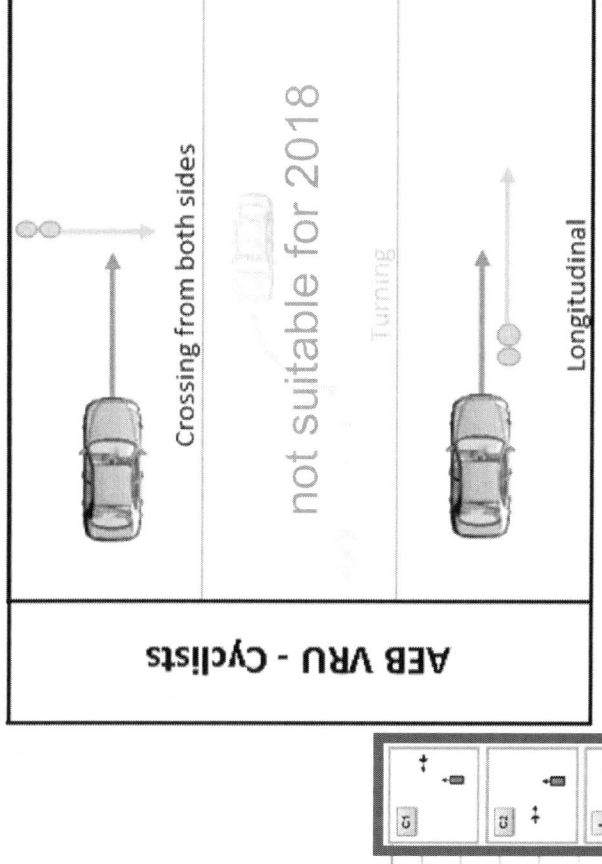

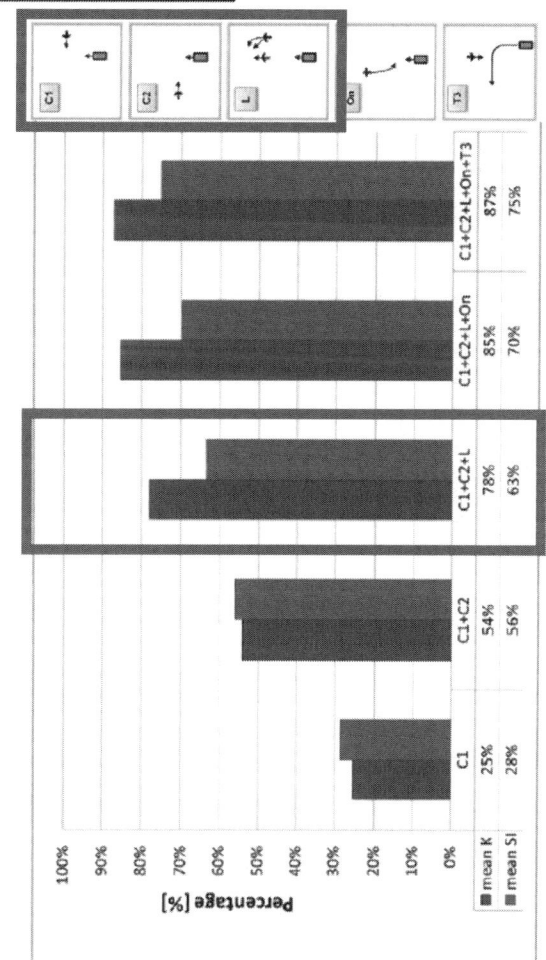

Source: CATS consortium: * „Cyclist-AEB Testing System", www.TNO.nl/cats

April 2016   Zukünftige Tests von Fahrerassistenzsystemen im Verbraucherschutz; A. Rigling; ADAC e.V.   19

AEB VRU - Cyclists

- Cyclist Dummy

www.TNO.nl/cats

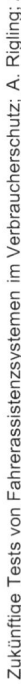

Lane Support Systems

- Potential updates 2018

  - Focus on LKA not LDW
    - no beeps, bings and bongs
    - effectiveness and acceptance

  - Road edge detection
  - Oncoming vehicles
  - Overtaking vehicles
    - active blind spot
  - Focus on straight roads

emergency
LKA

Summary

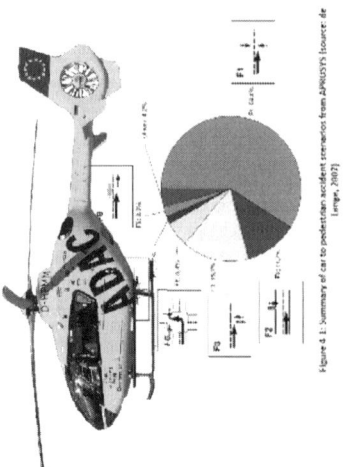

Key component dynamic motion controller – longitudinal and lateral vehicle control for automated driving functions

Dipl.-Ing. Benjamin Jüstel, Dipl.-Ing. Ulrich Stöckmann,
Fa. Continental Teves AG & Co. oHG, Frankfurt am Main

© Springer Fachmedien Wiesbaden GmbH, ein Teil von Springer Nature 2018
R. Isermann (Hrsg.), *Fahrerassistenzsysteme 2016*, Proceedings,
https://doi.org/10.1007/978-3-658-21444-9_5

1 Motivation

Automated driving not only represents the next step in the evolution of technology, it also provides added value in various ways, especially for the driver. Therefore the Continental strategy aims at automated driving to provide the driver additional benefits regarding relaxation, more time, enhanced safety and greater comfort and convenience. The development focuses on the way towards the vision of achieving zero accidents. Different use-cases like Traffic Jam Assist which supports the driver in dense traffic situations or Remote Garage Parking which allows the driver to delegate the maneuvering within private property to the vehicle are conceivable. The automated driving functions will change the individual mobility significantly. The technical evolution will be introduced step by step, depending on the participation of the driver starting with partially automated driving functions for low speed maneuvers and leading to fully automated driving on the highway (Figure 1).

Figure 1: Continental Roadmap Automated Driving

In addition to the technical requirements also legal and insurance aspects need to be covered for a successful introduction of automated driving. For example the Vienna

Convention on Road Traffic from 1968 which was revised in march 2014 to allow systems which influence the vehicle control when such systems can be overridden or switched off by the driver [2]. Involvement by all affected parties started on, so it is not the question whether automated driving is implemented, but when.

2 Functional Components and the Sense, Plan, Act Pattern

The required technical components for automated driving can be classified in the three main groups Sense, Plan and Act. Sense stands for components that are responsible for the perception of the vehicle state and its environment, such as sensors for pedestrian protection, wheel speed sensors, cameras, or short and long range radars. Plan is used for the derivation of possibilities for action and necessary decisions as they are made by various electronic control units. Act means a direct intervention in the control of the vehicle e.g. the brake caliper, the electro-hydraulic brake MK C1 or electric parking brakes. The Sense, Plan and Act control chain is derived from the tasks of a human driver who is nowadays the link between the brake, steer and engine actuator and the environmental sensing. Transferred to an abstract vehicle architecture the Sense, Plan and Act pattern results to the following logical components (Figure 2).

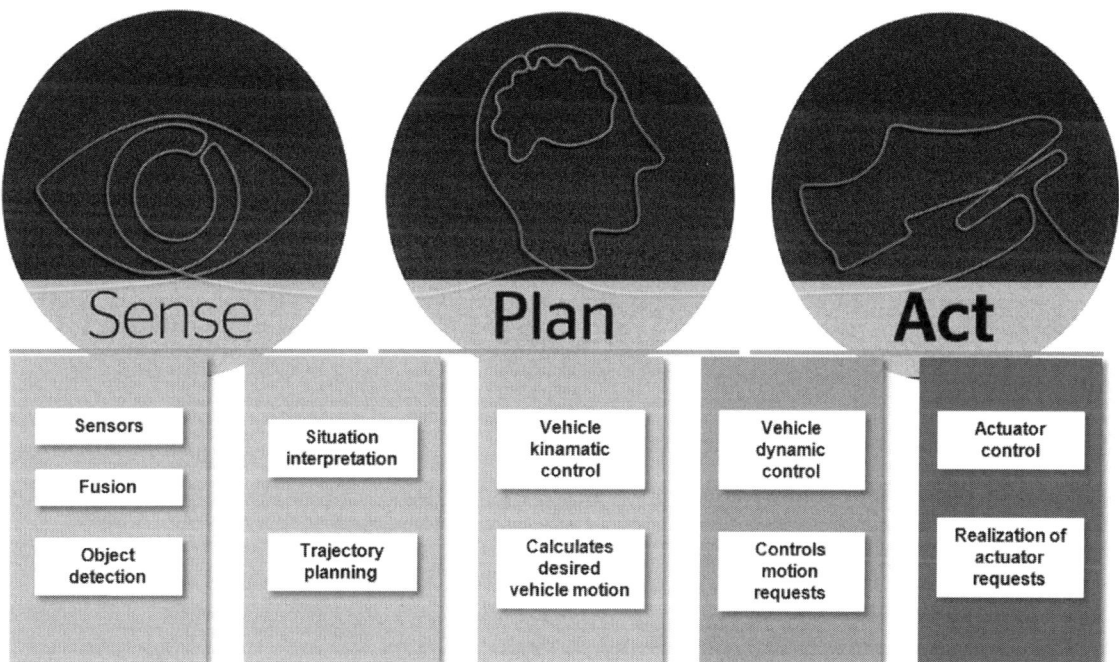

Figure 2: Sense, Plan and Act – Functional Components

Shown in the left area the vehicle state estimation based on the environmental sensing, e.g. object detection and algorithms for sensor data fusion. Resulting from the virtual environment model, the function characteristic and the situation interpretation a desired vehicle trajectory is calculated. The kinematic controller calculates the motion requests from the desired trajectory and the Dynamic Motion Controller realizes the request with the help of the available actuators. A detailed view on the Dynamic Motion Controller shows the following functional components (Figure 3):

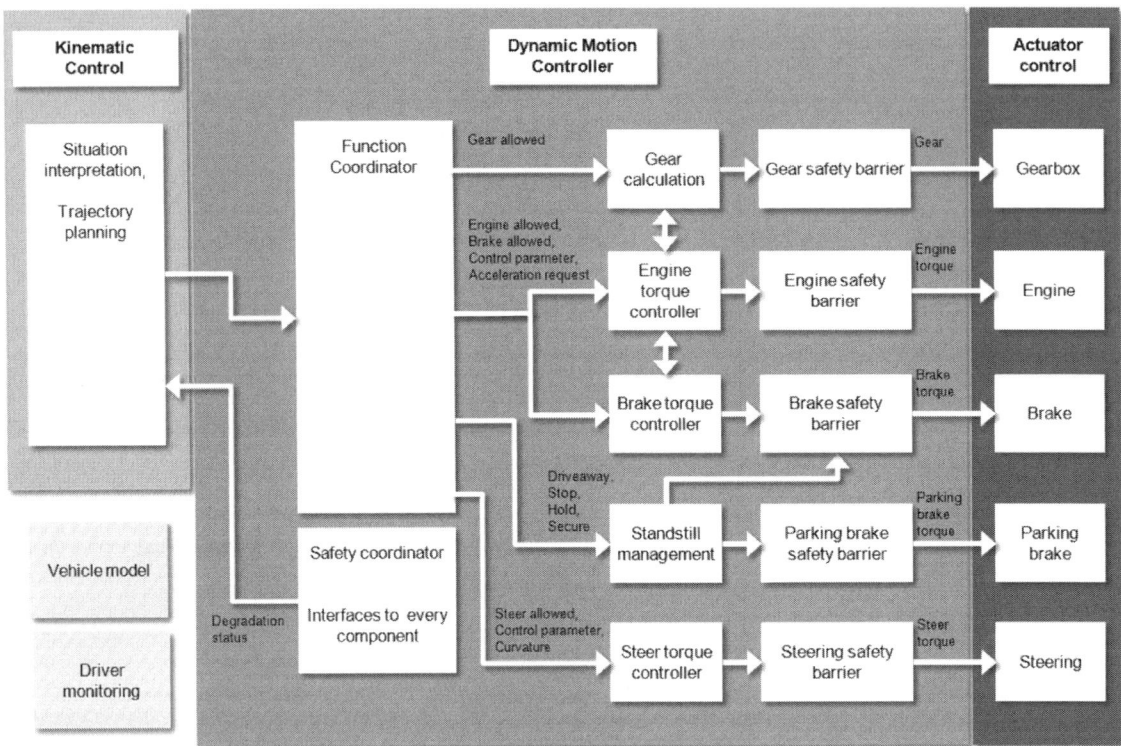

Figure 3: Dynamic Motion Controller – Functional Components

The function coordinator takes all requests from the kinematic controller into account and controls the torque controllers for engine, brake and steering. The safety barriers contain all mechanisms which are derived from the hazard analysis and the resulting functional and technical safety concept. The design of the safety barrier is depending on the function characteristic and includes e.g. a limitation of the absolute value or a gradient limitation. The safety coordinator monitors all components which are required by the function and is one important input for the safety barriers. The Dynamic Motion Controller is thereby the link between the main tasks of Plan and Act.

The functional architecture can be transferred to a technical architecture by partitioning to control units (figure 4):

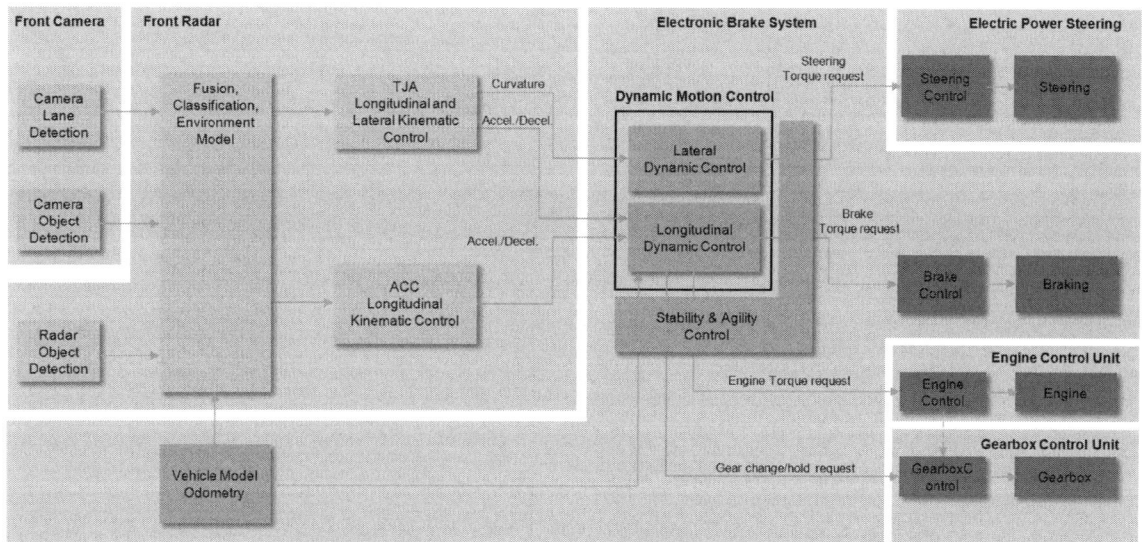

Figure 4: Technical Architecture – Function example ACC and TJA

In the exemplary technical architecture the camera and the radar sensor detect the objects and in addition the camera detects the lane markers. The kinematic controller which calculates the motion requests for the functions Adaptive Cruise Control (ACC) and Traffic Jam Assist (TJA) is located in the radar control unit. The Adaptive Cruise Control function controls the distance to the preceding vehicle and the driver does not need to apply the brake or throttle. If there is no preceding vehicle the function controls the vehicle speed to the target speed entered by the driver. The Traffic Jam Assist is a partially automated function i.e., the driver has to take over the control immediately if necessary. If a driver comes in a traffic jam situation and the vehicle speed is low the function will keep the vehicle in the center of the road and control the vehicle speed resp. the distance to the preceding vehicle until the end of the traffic jam. The Traffic Jam Assist function is so to speak the Adaptive Cruise Control part for the longitudinal vehicle control combined with a lateral vehicle control using the steering actuator. Taking over the lateral control part grants a significant relaxation for the driver (Figure 5).

Figure 5: Traffic Jam Assist

The interface between the kinematic controller and the Dynamic Motion Controller mainly consists of a curvature request for the lateral control part and an acceleration request for the longitudinal control. The output of the Dynamic Motion Controller provides the torque requests for the actuators engine, brake and steering. In the exemplary technical architecture the Dynamic Motion Controller is implemented in the electronic brake system which has some advantages, e.g. latency free access to secured internal signals, a central location for safety mechanisms as well as controller performance and the electronic brake system is usually included in low and high line platforms. The Dynamic Motion Controller torque request is secured by the stability control functions ABS and ESC which can modify the requests to grant vehicle stability and are therefore one important part of the safety barriers shown in figure 3. In addition to the engine, brake and steering request the Dynamic Motion Controller has also an interface to the gearbox e.g. to secure the vehicle in standstill or to prevent oscillation shift requests [3]. The longitudinal Dynamic Motion Controller includes the functional components gear request calculation, engine torque controller and brake torque controller shown in figure 3. The lateral Dynamic Motion Controller correlates with the steer torque controller. Taking the requests from the Dynamic Motion Controller into account the actuators are the last link in the chain and responsible for the transmission of the torque requests to the road.

The Dynamic Motion Controller is one key component in the Sense, Plan and Act chain and therefore mandatory for automated driving functions.

3 Key Component Dynamic Motion Controller

Nearly all automated driving functions require a component which controls a virtual driver request to drive a defined trajectory respectively safely realize a function motion request. This fact leads to the development of a central Dynamic Motion Controller to reduce complexity and save research and development costs. Due to the fact that many new vehicle functions will be introduced by the megatrend of automated driving the implementation of decentralized function specific controllers is not recommended.

The centralization of the Dynamic Motion Controller for different function use cases is the most challenging aspect for the development because of the broad variety of functional requirements. For example an emergency brake assist function or a road departure protection function will focus on a fast vehicle response on the other hand a traffic jam assist function will set priorities to a comfortable realization to ensure relaxed driving. The parameter application for the lateral and longitudinal Dynamic Motion Controller therefore needs to be function specific or at least function group specific. Furthermore the processing of the Dynamic Motion Controller torque output could be adapted by the function.

Another important requirement for the Dynamic Motion Controller beside the centralization is a modular structure to ensure high flexibility. The longitudinal and lateral control part should be configurable and not depend on each other.

4 List of abbreviations

ABS   Anti-lock Brake System

ACC   Adaptive Cruise Control

ESC   Electronic Stability Control

TJA   Traffic Jam Assist

5 List of literature

[1] SAE J3016, Taxonomy and Definitions for Terms Related to On-Road Motor Vehicle Automated Driving Systems

[2] ECE/TRANS/WP.1/145 – Report of the 68th session (24-26 March 2014)

[3] Handbuch Fahrerassistenzsysteme, Vieweg+Teubner Verlag, 2012

Next generation Opel Eye – mono-camera-based emergency braking assistant systems

Dipl.-Ing. Gernot Wiese
Dr.-Ing. Alexander Weitzel
Dipl.-Ing. Angelo Raczek
Adam Opel AG, Rüsselsheim

© Springer Fachmedien Wiesbaden GmbH, ein Teil von Springer Nature 2018
R. Isermann (Hrsg.), *Fahrerassistenzsysteme 2016*, Proceedings,
https://doi.org/10.1007/978-3-658-21444-9_6

1 Introduction[1]

Collision avoidance and impact mitigation functions for longitudinal traffic are playing a major role in the strategies to reduce accidents in public traffic. The potential of those assistant systems is already honored by the EuroNCAP consortium. Warning and autonomous braking features can contribute partial points to the overall vehicle star-rating [2].

The challenge for the automobile industry is to offer their customers efficient systems at affordable costs. One main cost driver in those systems are the sensors necessary to perform the functions. These sensors need to detect and classify potentially hazardous object in front of the vehicle, and provide data for a criticality evaluation of the situation. Depending on manufacturer and application, most of the functions currently in the market use RADAR or LIDAR sensors or fusion systems.

An alternative to those variants is a single sensor monocamera-based approach. One advantage of this concept is that various information supporting multiple features can be extracted from the captured optical data. With the launch of the new Opel Astra in 2015, the next generation of camera technology, called the "Opel Eye", is rolled out. It combines enhancements of known features like front collision warning (cf. [3]), lane departure warning, traffic sign recognition, and advanced exterior lighting assistant functions (cf. [4]), and complements it with the innovative lane keeping assistant, the glare-free high beam assistant IntelliLux LED®, and a monocamera-based emergency brake assistant. Combining twelve single features within one sensor with a single optical path pursues a cost-effective approach. However, the development of such a system to series-production readiness in a global company not only results in technical challenges, but also in questions about specification, organization, and development methods allowing the execution in globally distributed development teams.

Since launch of the Opel Astra K, customers can benefit from the emergency braking assistant system with the mono-camera. Subsequently, vehicles equipped with this system are rewarded with lower insurance rates.

---

[1] This paper is based on [1] and has been reworked and enhanced to cover most recent results

2 Challenges of Global Product Development

Besides the technical challenges, a factor of success for an efficient development process within a global company is mainly the interaction of the resources available in various locations. That concerns the teams of experts, for example for requirements specification to components, systems, and functions. But it also concerns the availability of test equipment, as well as the verification and validation on vehicle level. Additionally, the assignment of global teams allows the consideration of traffic behavior and scenarios specific to various continents and cultures. At the same time, the utilization of identical or similar systems in as many product of the global company as possible can improve the product cost.

Hence, functional specification and behavioral definition are formulated universally and independent of architectures and systems. To be able to cover regional aspects, means for configuration and parameterization are provided. The regional team utilize those, to consider legal differences and regionally specific consumer metrics.

A core development team performs the development of the base application, according to the universal specification, at one location, which is responsible for the first introduction of the feature in the first platform. This approach ensures a unified interpretation of the requirements.

The functional validation is distributed among the regional teams. They transfer the stage of base development to the different vehicles on various vehicle platforms and perform the integration tasks. That enables the identification of errors and deficiencies and in parallel allows further refinement of specific regional requirements.

Figure 1 shows schematically the process flow for the development of a Driver Assistance function in a global application.

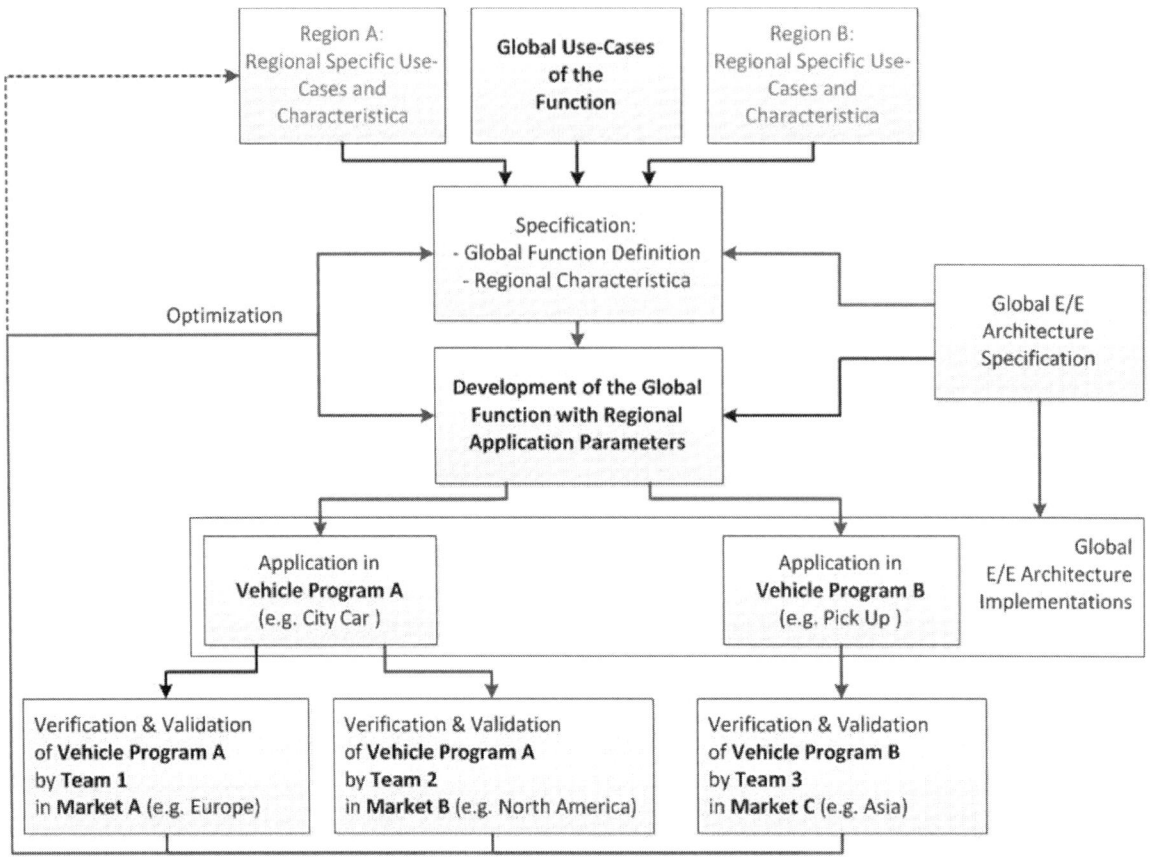

Figure 1: Functional development and optimization for a global function

3 Development Process

The development of active safety systems for global vehicle platforms requires likewise globally aligned processes, in order to accommodate the variety of programs and variance between regions and cultures. Additionally, the active safety functions need to be applicable to several systems with different electric/electronic architectures.

In an approach based on the V-model-process [5], the universal behavioral requirements of the active safety functions are detailed and substantiated on system level. The allocation of functional elements to subsystems and (if applicable, existing) components lead to the derivation of interface requirements for the participating control modules, and to the realization of the complete system within the given electric-electronic architecture. To allow this, the global architecture has to provide the flexibility to apply existing interface concepts to other system variants.

As a good example of this method, the existing RADAR-camera-based emergency braking system provided the template for the extension of the front camera-based system to support intervention strategies for longitudinal dynamics. The required interfaces for the brake and propulsion systems and the driver interaction were mostly already existent and mature. At the same time, this approach means that additional interfaces cannot be integrated later at all, or only with great effort.

Hence, but also for packaging and cost reasons, visual sensor and control module were integrated for the emergency brake assist function. This front camera unit is not only responsible for the object detection and classification, but also for the evaluation of the object data and the determination of the appropriate reaction. It generates the activation requests for the brake system, the engine control, and the driver information module, which execute the reaction and warn and alert the driver.

In parallel to the functional development process, an examination of the functional safety according to ISO26262 [6] takes place. The functional safety concept with its subsequent requirements is derived by a hazard analysis. For the camera-based emergency braking system, this results in the necessity of primary and secondary monitoring functions and plausibility checks (decomposition). The primary supervision and the required plausibility checks are contained in the camera. Since an erroneous activation of the autonomous braking is considered the highest potential hazard, a secondary monitoring in the brake system is required. It verifies the brake request by the camera with the present driving scenario before activating the brakes and it can, if necessary, prevent an unwanted or even unintended autonomous braking. All relevant interfaces in this loop need to be protected.

4 Technical Challenges

All previous Opel Eye generations used monocular camera technology. This kind of sensor technology for scanning the environment contains a single optical path, with one optical lens and one image processor.

The advantage of a mono-camera system for the vehicle integration is that only one sensor is required. In order to achieve a large viewing range starting right in front of the vehicle with a given opening angle, and to benefit from already existing cleaning functions, the sensor is located behind the windshield close to the base of the rear view mirror. Thereby, the sensor is part of the vehicle interior, which reduces the requirements in terms of dirt and moisture. Stains in the viewing area of the camera can be cleaned by the windshield wipers. Disadvantages of this packing location are potential limitation of the driver's field of view and the interference with other components in this area, such as the rain-light-sensor.

Another advantage of the front camera is the similarity of this sensor's way to scan the environment with the human visual perception. Thus, the camera is able to access a variety of available visual information in traffic, such as lane boundaries, traffic signs, traffic lights, etc., if it is capable of extracting these characteristics from the captured images.

The usage of a single optical path for scanning the environment has also disadvantages. For the mono-camera these are especially the dependency on weather conditions (like rain, snow, fog, etc.), limitations in twilight (limited to the area lit by the head lamps) but also in extreme bright conditions (sun glare), and shorter range compared to the RADAR sensor. Limitations due to weather conditions can be detected by the camera itself and it allows to restrict the feature functions accordingly. The integration of further functions into the camera increases the required computation power, resulting in the risk of increasing product costs. However for the new generation camera, the product costs could almost be kept unchanged compared to the previous generation, despite a larger functional content.

With lower costs, this solution allows the usage of the system in a broader variety of vehicles, especially in the very cost-sensitive segments. In addition, the system and requirements definition provides a scalability, which enables the integration of the mono-camera into a sensor network that is applicable to vehicles with higher performance requirements.

The specific challenge for a mono-camera for a variety of driver assistant functions is the reliable extraction of multiple information form the visual image. This image processing in the mono-camera takes three processing steps. First, the image is searched for specified forms and patterns. Recognized characteristics are grouped and combined to objects. The identification and classification of the objects results from the entity of all recognized

6

characteristics for an object. The objects are divided into classes, like traffic signs, lane markings, vehicles, or pedestrians. At the end, specific object information, like relative distance, is derived from single images or sequences of images. The relative distance to a preceding vehicle, for example, is determined from the object size and the position of the object relative to the camera's virtual horizon [7]. The relative velocity or the trajectory path of an object, though, is derived from the change of position and size of the object in a sequence of images [8].

Beside these dynamic data, the emergency brake assistant function requires the determination of the Time-to-Collision (TTC). It describes the time remaining to a potential collision. When the TTC during a driving scenario falls below a certain threshold, the feature warns the driver with a visual and acoustic alarm. Additionally, the brake system is prefilled and the thresholds of the panic brake assist function are lowered. If the driver still does not react to the warning with a proper braking or evasive maneuver, the emergency brake assist triggers an autonomous emergency braking. Depending on the actual driving scenario, lighting conditions, viewing range, and weather conditions, the system might avoid a collision up to 40 km/h and mitigates the consequences of the collision up to 80 km/h. It offers dynamic brake support in the entire operating speed range to boost the driver's effort of braking even if the conditions for an autonomous activation are not met. With that, typical urban traffic scenarios are covered.

The operating speed range for the feature of up to 80 km/h has been chosen to suit the sensor's capabilities. Further extension would have been possible but only at the costs to increase the risk of false activations. These again would inflict the customer acceptance, especially in city traffic situations that have been the focus of the whole implementation. The limitation of maximum autonomous braking deceleration to 40 km/h is also contributed by the system safety hazard analysis. A single-sensor system could not be granted a higher braking authority, unless the driver confirms the threat of a collision. In this case, the system supports the driver's attempt with maximum deceleration in the entire operating speed range. After all, the current accident statistics for Germany show that about 73% of the accident happen in city traffic [9].

5 Verification and Validation

The usage of a mono-camera system in a global variety of vehicles sets high standards for development and verification processes, due to missing redundancy in object detection. Additionally, requirements of various regions with specific use cases need to be considered. At the same time, the increasing number of relevant scenarios and test cases, and thus the respective validation effort, can be directly used to drive the quality of verification.

Due to the direct control vehicle dynamics by the emergency braking function, all functional safety aspects need to be considered and validated with appropriate methods. System boundaries and sensor limitations, which might lead to inappropriate or unintended activations of the system, need to be taken into account.

The use cases for the collision, are defined by requirements derived from accident occurrences and test protocols, e.g. by consumer metrics. Due to the risk of collision, the use case validation must be carried out on proving grounds with suitable target objects.

Apart the justifiable collision avoiding activations, which the customer should only experience very rarely, those activations that the driver perceives as unnecessary require attention. They have a large potential of lowering the customer satisfaction and might reduce the experienced product quality as well as the customer acceptance for the system. Many unwanted activations of the system could drive the customer to deactivate the feature and, hence, eliminate the benefit.

In order to minimize these cases, a large amount of mileage with the system in typical driving conditions and in many different driving scenarios needed to be captured. The effort benefitted from the limitation of the system's operating speed. In addition, mechanisms to speed up the activity could be utilized by focusing on these accumulation scenarios that pose the highest performance requirements on the object detection and evaluation capability of the system. For this collision avoidance and mitigation function, containing warning and braking intervention, the preferred situations were in city traffic with a variety of moving and stationary objects on diversified trajectories. To increase the risk of unwanted (respectively unintended) activations, in most of these situations, the driver would be highly attentive (Figure 2) and aware of the situation.

Several influencing factors have to collude, so that an activation is triggered inappropriately in a certain scenario. Therefore, and since these can occur in complex driving situations, the reproducibility of single scenarios is small. That creates the following challenges for the development.

In the potentially small number of actually experienced driving situations it is crucial to identify patterns and characteristics that reveal weaknesses of the system. In that process, a local optimization of a single system property at the expense of other scenarios must be

Figure 2: Driving scenario with several captured objects

avoided. Experiences in many different international traffic situations provide the chance to support this identification, since a wider range of potential scenarios is covered. At the same time, this wide spectrum requires to identify market-specific characteristics to address regional specific application in the development of the functionality (cf. Figure 3).

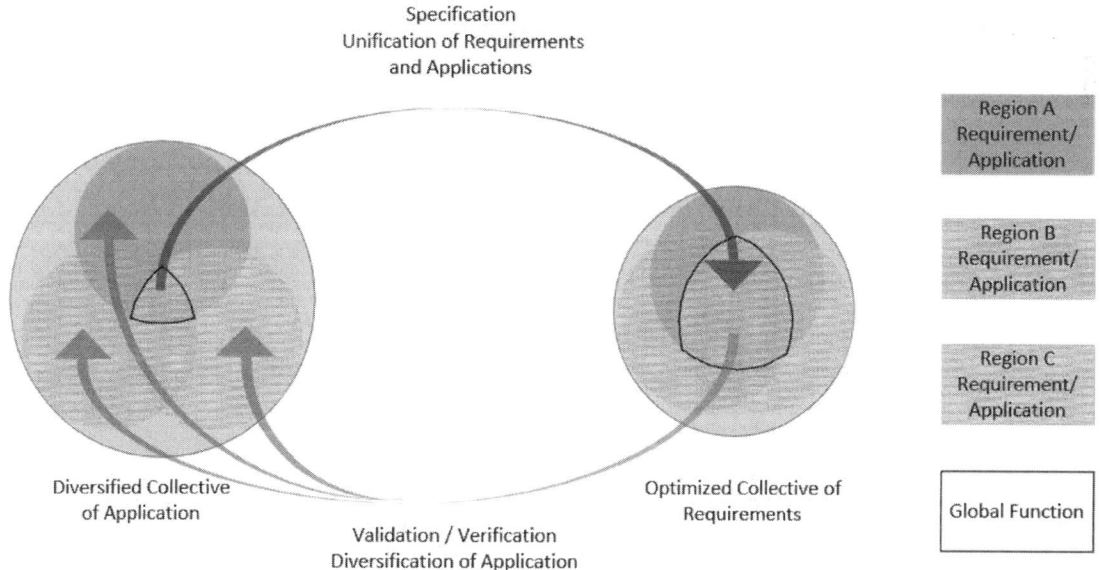

Figure 3: Conflicting Interest: Regional Application vs. Global Requirements

Another challenge after an optimization loop of the function is the need to re-run the scenarios that were identified as critical, in order to proof the effectiveness of the optimization. Due to the limited reproducibility of scenarios in real-world driving, Hardware-in-the-Loop (HIL) methods re-simulating those critical situations are a key enabler.

6 Evaluation of Benefit

The Emergency Braking Assistant System described in here is an enhancement of the Forward Collision Alert system that has been already available in series production for several years [3]. As a warning feature needs to act earlier to address the reaction time of the driver, more frequent activations of the alert are expected compared to autonomous emergency braking. Based on the existing feedback out of the field and the available validation results, a very low rate of false activations can be assumed. A recent cross-manufacturer study on emergency braking systems projects a potential reduction of rear-end collisions of 38% [10]. As a side effect, other active safety features integrated in the front camera have the potential to lower the driver workload, and thereby enhances the possibility to pay attention to and concentrate on the current driving situation.

The Emergency Braking Assistant System contributes to the overall EuroNCAP 5-star rating for the Opel Astra K [11]. In addition to the well-received crash ratings, the vehicle is rewarded with a lower insurance rate if equipped with this system. Hence, the customer benefits in multiple ways from opting into the Emergency Braking Assistant System.

To further increase the take rates by enabling a positive customer product perception, the front camera system allows the driver to experience some of the features they pay for. As they should never be forced to benefit from the Emergency Braking Assistant, additional features, especially the Following Distance Indicator (and in some cases the Forward Collision Alert), deliver a feedback in normal driving situations. They show a preceding vehicle being detected and tracked by the camera sensor and provide the current time gap on demand in the cluster. In addition, there is feedback by the "vehicle ahead" telltale illuminated constantly if a vehicle is detected and changing from green to amber at small remaining time gaps. This telltale enhance the known warning and braking cascade and is a first indicator for the driver which can be experienced in daily driving situations, without being a nuisance. It also gives the customer a chance to learn limits of the system (e.g. cut-in maneuvers, range limitations in curves, etc.). In the warning cascade, the telltale is followed by an acoustic alert and a visual stimulus with red LED´s mirrored in the lower windscreen. These are replicating flashing brake lights and are able to quickly redirect the drivers focus to the situation ahead (cf. Figure 4).

Figure 4: Replicating flashing reflexive LED warning light

If the driver, despite these heads-ups, still does not act appropriately the Emergency Braking Assistant System will activate an autonomous braking.

Overall, the driver should not experience the autonomous emergency braking very often in a vehicle's life-time. An internal study in the North American market shows that an emergency braking activation occurs way less than once a year per vehicle (heavily depending on driver type, driving profile, and location factors). If the driver is rarely experiencing the benefits of the feature, unwanted activations could be perceived as even more severe and need to be avoided. But how large is the risk of annoyance or even false activation?

Data captured during the development and the validation of the emergency braking feature, by both the OEM and the suppliers, show that the driver was actually assisted by the feature in several scenarios that might have led to a collision. On the other side the analysis of the final validation data accumulation and the re-simulation revealed no false positive activation. This is backed up by internal and external driver feedback reports not showing incidents of unwanted braking.

However such results does not necessarily allow to conclude a valid statement that false positives cannot occur. The validity heavily depends on the mileage driven with test fleets to enable any statistical significant analysis. To prove even a conservative false activation ratio could require millions of kilometers for a robust analysis [12]. As described before, a chance to limit the effort lies in the balancing of the targeted use cases versus the sensor limitations and the validation collective.

Additional support of this validation dilemma is possible due to the sensor concept of picture perception in combination with external data links. In conjunction with the identification of parameters for the filtering of potential false positives this is a key enabler for development of features for low occurrence use-cases.

7 Conclusion and Outlook

The monocamera-based Emergency Braking Assistant System represents a consequent enhancement of camera-based driver assistant systems that are already offered by Adam Opel AG and General Motors Corp. To utilize the full potential of this function in terms of product costs as well as the strategy for avoiding accidents, the roll-out into as many global vehicle variants as possible should be pursued. This requires the consideration of regional specifics while maximizing the globally uniform aspects. The product development processes and the used system architectures need to cover these correlations but still allow for compromises that are necessary in the realization of complex functionalities. At the same time, this approach enables globally distributed teams to work in parallel in verification and validation. Their results and experiences can be used for enhancement of product definition as well as execution. The permanent differentiation of regional influences and common weaknesses of the system are both a blessing and a curse. Those require a close and constructive work relationship of the international team. That might be more time consuming, but allows the consideration of many different driving scenarios and use cases. It also generates a systematic redundancy for critical situations. To tackle the overall challenge of validation of active safety features in low occurrence real life traffic situations the camera offers particular benefits.

In the global system network, the concept of a mono-camera with the combination of image processor and application control module has proven to be a flexible and cost-effective solution, even in the next stage of extension. The potential range of applications are not yet exhausted, further functionalities based on this sensor concept are possible. The relatively large number of assistant functions, which could be realized with this single sensor, allow even in smaller vehicle platforms the introduction of functions that are normally limited to upper-class vehicles. That provides the opportunity to offer low priced comfort features in the cost-sensitive vehicle segment. Thereby, all participants benefit from a safer traffic conditions.

Bibliography

[1] Weitzel, A.; Raczek, A.; Wiese, G.: Next Generation Opel Eye – Mono-Kamera basierte Notbremsassistenzsysteme, Konferenz Elektronik im Kraftfahrzeug, 14.-15.10.2015; Baden-Baden, 2015

[2] EUROPEAN NEW CAR ASSESSMENT PROGRAMME ASSESSMENT PROTOCOL (Euro NCAP) – SAFETY ASSIST, Version 6.0 July 2013; available at: http://euroncap.blob.core.windows.net/media/1572/euro-ncap-assessment-protocol-sa-v60.pdf; fetch on: 22.06.2015

[3] Raphael, E.; Kiefer, R.: Development of a Camera-Based Forward Collision Alert System, SAE 2011-01-0579, 2011

[4] Der neue Opel Astra, ATZ Band 111 Ausgabe 12, 2009 S. 888-891

[5] V-Modell XT; available at: http://www.cio.bund.de/Web/DE/Architekturen-und-Standards/V-Modell-XT/vmodell_xt_node.html ; fetch on: 22.06.2015

[6] ISO 26262 – Road vehicles – Functional safety, International Standardization Organization, 2011

[7] Whydell, A.; Heinrichs-Bartscher, S.: Vergleich von Mono- und Stereokameras, in: Automobil Elektronik (2014), Nr. 3, S. 24 – 27.

[8] Lee, D. N. (1980): The optic flow field: the foundation of vision, JSTOR. Philosophical Transactions of the Royal Society of London – Series B: Biological Sciences, 290(1038), 169-79.

[9] Statistisches Bundesamt (Hrsg.): Verkehrsunfälle 2014. Fachserie 8, Reihe 7, Artikelnummer: 2080700147004, Wiesbaden 2014

[10] Fildes, B.; Keall, M.; et al.: Effectiveness of low speed autonomous emergency braking in real-world rear-end crashes, Accident Analysis and Prevention 81; S.24–29, 2015

[11] EUROPEAN NEW CAR ASSESSMENT PROGRAMME (Euro NCAP) - TEST RESULTS of Opel/Vauxhall Astra 2015 available at: http://euroncap.blob.core.windows.net/media/22086/euroncap-2015-opel-vauxhall-astra-datasheet.pdf; fetch on 22.02.2015

[12] Winner, H.: Quo vadis, FAS?. In: Winner, H.; Hakuli, S.; Lotz, F.; Singer, C.: Handbuch Fahrerassistenzsysteme, 3. Auflage, Springer Vieweg, Wiesbaden, 2015

Stress-free parking –
thanks to a personal parking assistant

Heiko Herchet

Elina Schäfer

Alexander Süssemilch

trive.me – an EDAG Engineering GmbH trademark

1 Introduction

Valet parking is a service that relieves the driver of the bother of finding a parking space. Originally appropriate personnel (a valet) took over the vehicle, e.g. in front of a car park, hotel or at a restaurant and parked the vehicle. When the vehicle user wanted to resume his trip, the vehicle was turned over to him at a defined location.

Valet parking also refers to automated systems with which vehicles are autonomously taken to parking spaces by an automatic machine, such as a vehicle carriage or rail system. In particular, this technology is used in areas where the space required for vehicles must be optimized or minimized. In addition to saving the driver time and relieving the driver of stress, through valet parking the infrastructure (road, car park area) is also used more efficiently.

The parking process is intrinsically still a special challenge today and the complexity involved in parking, particularly in parking garages, is increasingly significant. To a great extent this complexity is due to the size of parking garages or the congested infrastructure. In a large parking garage, finding your car (again) after you have parked it, quickly becomes a problem.

Moreover, modern assistance systems in parking garages are barely functional.

The reason, for example, is the inaccuracy of vehicle location via GPS, as used in navigation systems, consequently it is difficult to make driving decisions within a parking garage. Different parking levels and a lack of wireless connections are additional aggravations.

Efficient use of the available parking space is influenced negatively by the self-determination of the vehicle driver and his driving skills.

A potential solution is an assistance system, which through navigation within the parking garage, supports the driver by finding a free parking space, and through direct settlement. In the first step, the functionality can be implemented via an app on the user's mobile phone (iOS, Android). The parking garage service, improved through this solution, offers an advantage for the user, and it also offers an advantage to the car park provider – thanks to the assistance system the attractiveness and the efficiency of his parking garage increase. These thoughts were the background for the development of the service, "trive.park – your personal parking assistant", which is described below in detail.

2 Context

Digitalization of the world continues unabated and is a self-evident support for our daily routine. Systems that require users to forego the conveniences offered through networking are finding less and less acceptance. The car is a "digital latecomer" in this world. trive.me is committed to this digital transformation and develops its own software services for Driving 4.0.

Unlike the consumer sector and the industrial sector, where networking is now an indispensable development driver, digitalization is still progressing at a very slow pace in the automotive sector. In this regard, the networking of all systems may ultimately not only be an advantage for the vehicle's range of functions; it may also be an advantage for mobility per se. Networking makes it possible to obtain information necessary for driving recommendations and for automated processes in the vehicle, from an external database. This fact opens up entirely new opportunities for the development and evolution of driving. Because the larger the database is, the more specifically it can be used.

Bringing the best of the automotive industry and software thinking together

For automotive development to remain marketable in digital competition, there is no alternative to bringing together the best work methods and thinking of the automotive and the IT industry. trive.me is an independent EDAG Engineering GmbH brand that specializes in developing digital solutions for networking the vehicle, driver, and infrastructure. At trive.me the EDAG know-how gained from complete vehicle and production plant engineering is incorporated into new, integrated approaches and solutions that are aimed at sustainably changing mobility in light of "Driving 4.0" to the same extent as the introduction of apps did for smartphones.

In the area of highly automated driving, there are still challenges that have not yet been solved, for example, there are various legal issues or the algorithm programming for decisions in borderline ethical cases. This is why intelligent parking in semi-public areas, such as parking garages, is one of the first steps towards highly automated driving.

High-level automation is only possible through networking

However, such a highly-automated parking process in parking garages is only possible with the networking of vehicle and infrastructure. For example, a connection to the parking guidance system of cities is necessary to obtain information concerning parking garages with available space. Likewise, user requirements should influence the selection of a suitable parking space. Thus it is ensured that an electric vehicle can also be positioned over a loading coil, or that the packaging in parking garages can be enhanced by taking vehicle dimensions into account. Other important points that do not function without this networking are the provision of a contactless access (e.g. automated barrier opening), integration of anautomatic payment process, and reliable route calculation to the next free parking space. These thoughts have been bundled and comprise the use case of the "trive.park" product described in section 3.

3 Idea and challenge

The service developed by trive.me bears the name "trive.park" – your personal parking assistant". "trive.park" precisely ensures the networking components that are necessary for an automated parking process.

The use case, presented in Fig. 1, describes this mission "Parking 4.0".

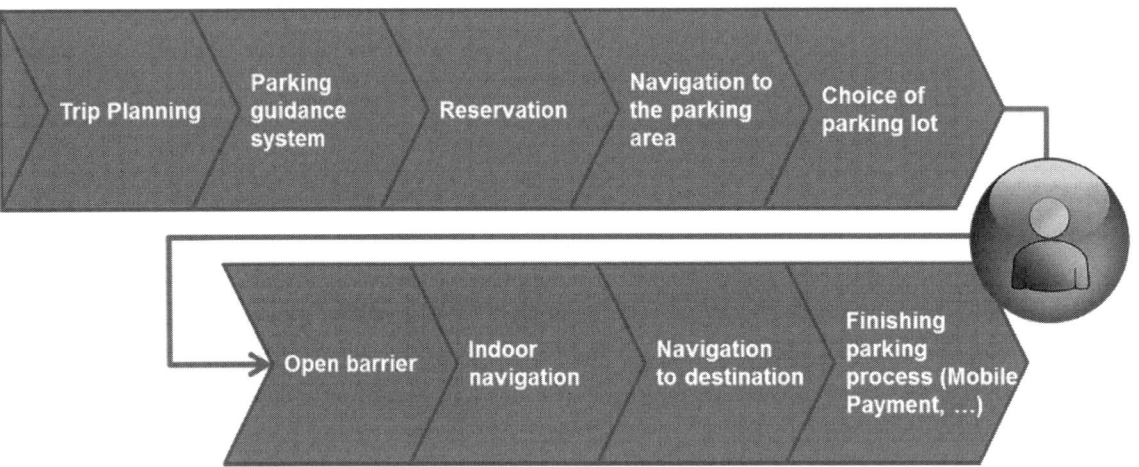

Fig. 1 Presentation of the specific sub-steps of the use case "Parking 4.0"

At the beginning of his trip, the user has the possibility of planning the journey or trip. Based on traffic information the time required is suggested and a suitable time for starting the trip is displayed. With the aid of the connection to the infrastructure, a suitable parking space is recommended through data exchange. For trips with an electric vehicle, in particular, it is conceivable that a reservation system for charging columns can also be implemented. In the next step the parking space is sought. The parking space is selected based on the vehicle and the user requirements. In addition, for off-street parking the barriers in parking garages are opened automatically and an (indoor) navigation system guides the vehicle user to the free parking space. Then navigation to the user's actual destination occurs. For the last step, the "settlement of the parking process|", the user is guided back to his car (CarFinder) and the parking fee is paid via "mobile payment".

One of the major challenges is fast implementation of this extensive use case. To do this, competencies of different companies will be bundled to enable fast implementation. In this regard, each company concentrates on its strengths. Together we map a consistent use case for maximum customer benefit and for an interesting application for vehicles.

4 Concept

However, in order to implement the idea of the overall use case, different components are required. These enablers are presented in Fig. 2 and are introduced in more detail below.

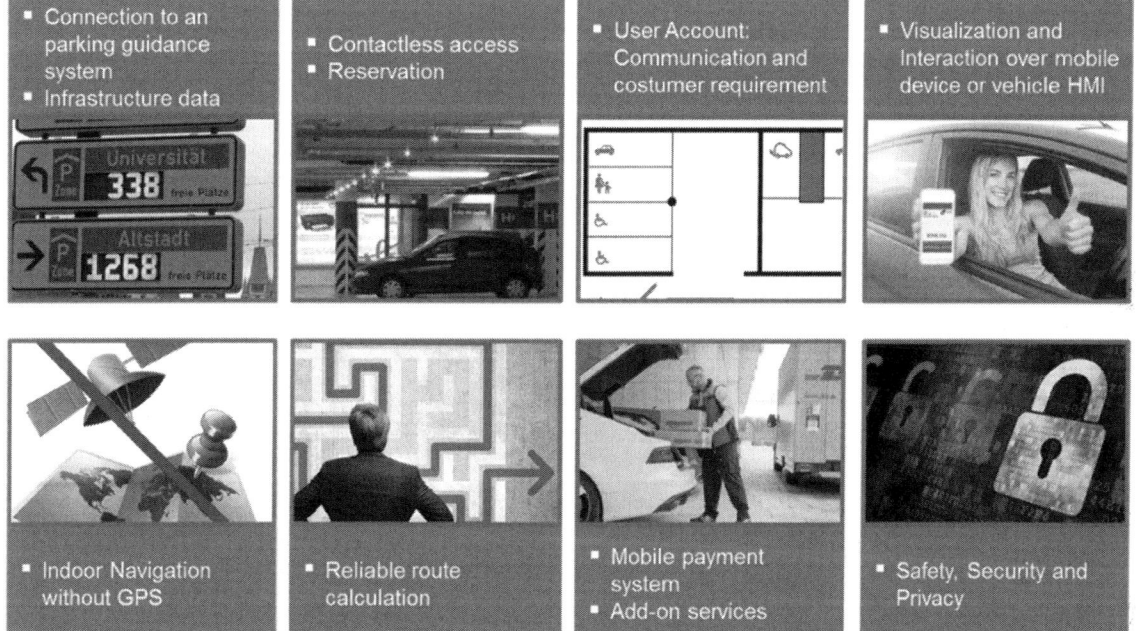

Fig. 2 Core topics – trive.park

1) Connection to a parking guidance system/infrastructure data

To obtain data concerning free parking areas, trive.park receives information from parking guidance systems.

Objective: Connection and a prognosis of free parking areas

2) Access to a free parking area/reservation

For the convenient parking process, it should be possible to reserve parking spaces in advance, and access to free parking areas should be implemented with convenient barrier or gate opening systems.

Objective: Opening barriers without a ticket

3) Customer profile: Communication – customer requirements

To meet the needs of customers, extras, such as information concerning the parking spaces, should be made available.

Objective: Rendering/filtering of information concerning the parking spaces, such as handicapped parking space, charging infrastructure, women's parking area, mother-child parking space, etc.

4) Visualization and interaction via mobile device or vehicle

The application should be visualized via mobile devices and the vehicle itself, and interaction should be possible.

Objective: User-friendliness and personalization

5 and 6) Reliable travel route calculation/indoor navigation without GPS

Reliable routing to a specific parking space should take place in areas where navigation has not been possible so far, such as parking garages.

Objectives:

- Navigation to parking garages or to the free parking areas
- Navigation in the parking garage to the booked/selected parking space
- Navigation from the parking space to a specific place
- Navigation from a specific place to the parked car (CarFinder)

7) Connection – payment system/supplemental services

To make the parking process more convenient, and to offer the customer a special service, in addition to a secure booking and payment system, supplemental services, such as auto care or parcel delivery, should also be implemented in trive.park.

Objectives:

- The payment process, possibly with discount systems
- A repair service/auto service
- Social media communication

8) Data security – secure communication channels

Because private data is exchanged in the communication between server and client, in addition to usability, the topic of data security also plays an important role.

Objective: Assurance of security, safety, and privacy

Overall technical concept – open interfaces as enablers for fast implementation of the complete use case

trive.park's overall technical concept includes different modules that each implement and map a component of the use case chain. This structure is presented in Fig. 3. Thanks to this modular structure it is possible to offer a minimal solution that functions in every parking garage and that can be specifically optimized as needed through local adaptations in the second step.

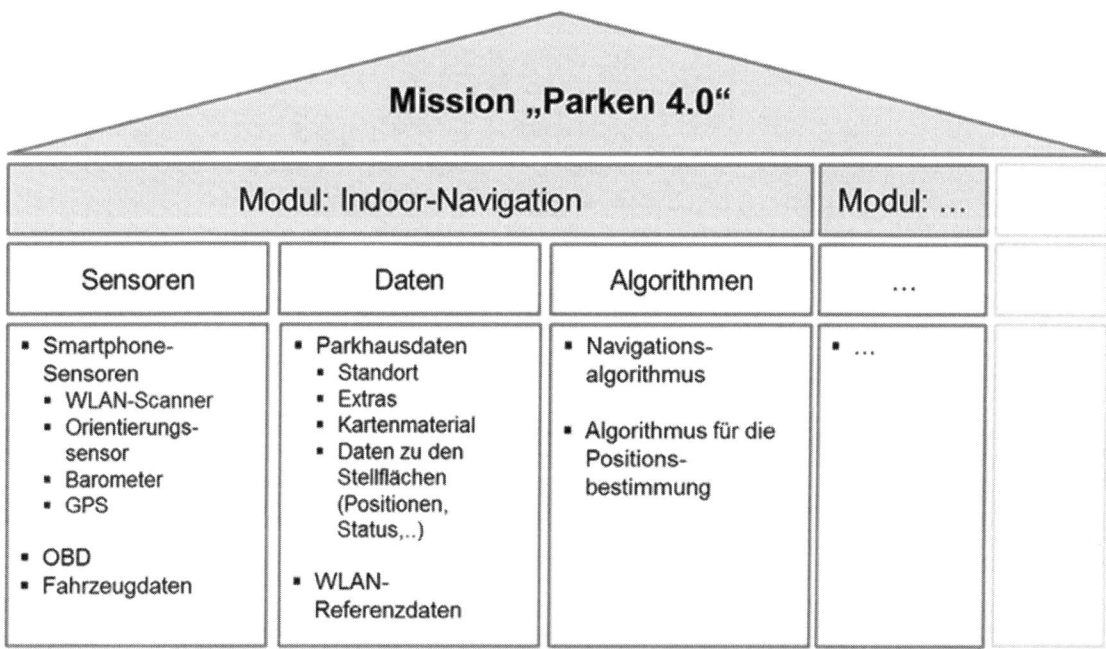

Fig. 3 Modular structure enables optimization of the service

The individual modules include the payment process, barrier opening, or indoor navigation. Networking with the infrastructure is assured through the connection to the Cloud. As a result, for example, the user gets information about free parking areas and can communicate bi-directionally with the parking garage. Thus it is possible to respond to the individual requirements of the user, and only visualize the data that is relevant for the user. For example, the owner of an electric vehicle obtains precisely the information that is applicable for free charging possibilities.

Indoor navigation is another technical focus, as GPS does not function in the indoor area. Various data sources, such as sensors from the vehicle, smartphone, and the infrastructure (e.g. Bluetooth LE) serve as a replacement that make it possible to determine the position with different positioning algorithms; optimally with an accuracy of 0.5 meters.

The highest premise in the development of the overall use case chain is IT security that ensures the necessary safety, security, and privacy for the implementation.

However, high user acceptance of the service will only be achieved if the service can also be used while driving. In the first step, the trive.park visualization occurred on the smartphone, however, in the future, the service will also be integrated in the vehicle. Currently this step is still difficult, because the standards such as "Mirrorlink", "Apple CarPlay" or "Android Auto", do not permit every third-party application. Moreover, it is conceivable that in the future these services will be integrated via separate SDKs from vehicle manufacturers. At the same time, through such a measure an exclusive access to selected vehicle data could be enabled to further optimize services.

5 Implementation

Two of the core topics cited in section 4 are described in more detail in this section. Additional information concerning implementation of the indoor navigation and the user account is provided.

5.1 Indoor navigation

Indoor navigation is a focus area for implementation of trive.park. The different concepts that are used for positioning in the indoor area are shown in Fig. 4.

The concepts for positioning function in isolation, however they can also be fused. Through the fusion of different positioning concepts locating accuracy can be increased.

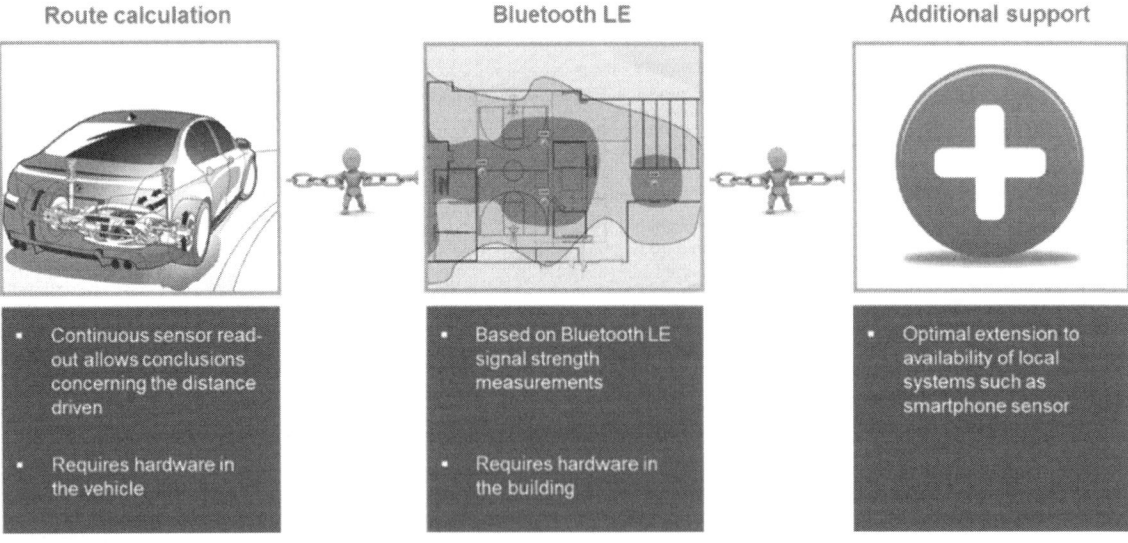

Fig. 4: Positioning concepts

Route calculation

For route calculation, continuous sensor read-out allows conclusions concerning the distance driven.

The following CAN messages must be used for route calculation:

- Momentary speed
- Steering angle of the front axle (effective)
- Inclination
- Yaw rate

In addition, the following vehicle data must be determined:

- Wheelbase
- Axle load distribution

The CAN messages cited above have different repetition rates. Vehicle position is recalculated on arrival of the message with the highest repetition rate. In this case this is the vehicle speed, its value is applied on the CAN bus every 20 ms.

The recorded speed can be broken down into its sub-components through the inclination of the road. This results in the vertical speed and horizontal speed that make a relative assessment of the elevation or the vertical distance within the parking garage possible.

The single-track model, shown in Fig. 5, can be used to calculate the curve radius of the vehicle's center of gravity with a given steering angle δ_L. The single-track model is used to explain the transverse dynamics of two-track motor vehicles, where both wheels of an axis are consolidated to one wheel. Lateral forces and side slip angle are disregarded.

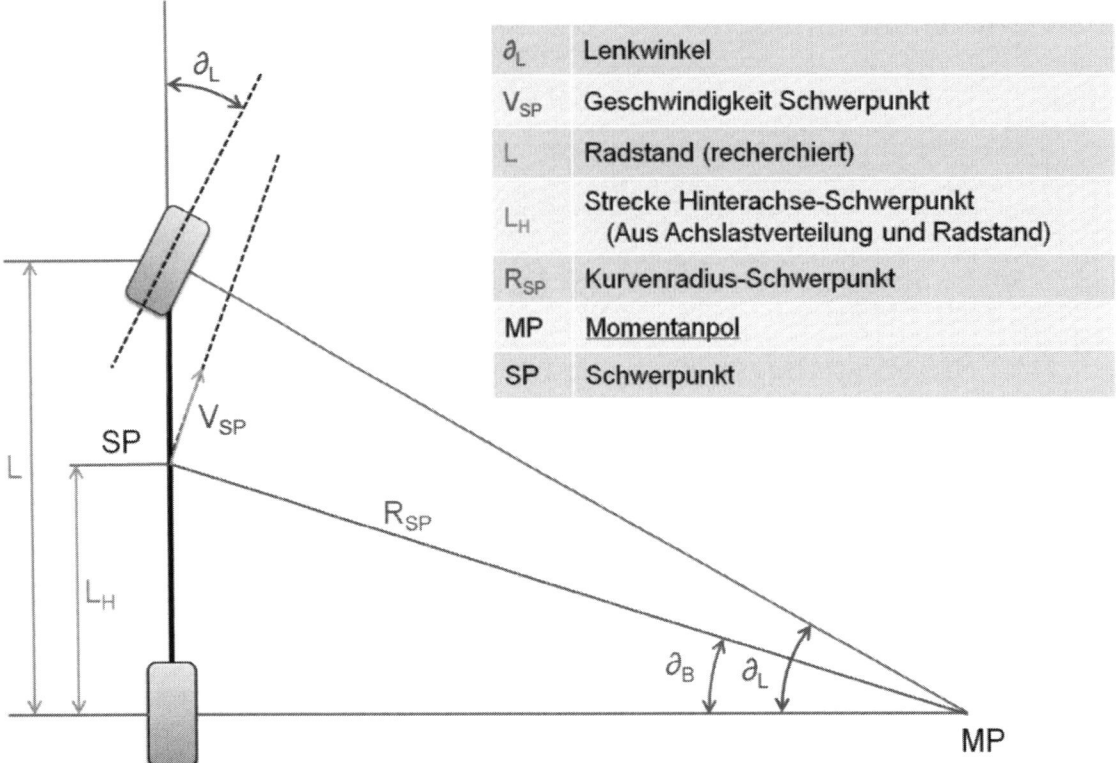

∂_L	Lenkwinkel
V_{SP}	Geschwindigkeit Schwerpunkt
L	Radstand (recherchiert)
L_H	Strecke Hinterachse-Schwerpunkt (Aus Achslastverteilung und Radstand)
R_{SP}	Kurvenradius-Schwerpunkt
MP	Momentanpol
SP	Schwerpunkt

Fig. 5: Single-track model

The curve radius RSP can be calculated as follows:

$$R_{SP} = \sqrt{\frac{L^2}{\tan^2(\delta_L)} + L_H^2}$$

A leg with the driven angle δ_{TSi} results from the read-out speed and the calculated curve radius.

The current vehicle orientation δ_{IST} is calculated by totaling all δ_{TSi}. In addition, the actual vehicle position can be calculated by totaling all Δx and Δy.

The simplified single-track model is ideal for slow driving around curves. For taking curves at fast speeds the yaw rate of the vehicle can also be included via an adaptive algorithm.

Bluetooth LE

For position determination via Bluetooth LE, the actual vehicle position can be derived from the signal strength patterns of the iBeacons in the vicinity. In the ideal case the Bluetooth signal strength decreases geometrically with increasing distance to the iBeacon. In reality the signal is subject to multipath propagation. Reflections, scatter, and refractions occur due to multipath propagation. These influences cause the signals of the iBeacon to reach the receiver through different paths and superposition effects, gain, and attenuation occur.

In an initial phase a data set with reference data will be collected for localization via fingerprinting. To do this, a signal strength measurement must be executed at different positions in the space, and the results must be stored in a database. Localization to the precise meter can only take place after this data set has been created.

Localization occurs in a second phase, in which a signal strength measurement is executed and the results are compared with the database. The received signal strength is used as characteristic signal, from which a vector for each fingerprint can be derived; the dimension of the vector is the number of received iBeacons. Euclidian distance is one of the simplest possibilities for determination of position. The similarity of two fingerprints can be calculated via the Euclidian distance.

Euclidian distance is defined as follows:

$$d(\vec{x}, \vec{y}) = \sqrt{\sum_{i=1}^{n} (x_i - y_i)^2}$$

Using this procedure, first a pre-selection of three fingerprints is made from the available data. Then the actual position is determined with even greater accuracy through trilateration. The distance between two reference data sets can be determined with the following formula:

$$P_r = P_0 \cdot d^\alpha$$

Where P_r is the reception power, P_0 is the transmission power, d is the distance between transmitter and receiver, and α is the path loss exponent. α can assumed in free space with 2, and in a building between 2 and 4. If you convert the formula for d and set the signal strength in the logarithmic dimension one, the following formula occurs:

$$d = \frac{10\log_{10} P_r - 10\log_{10} P_0}{-10\alpha}$$

For calculation of the position, three equations of circles are formed around the three closest neighbors with the calculated distances. As a rule, a point of intersection does not occur, due to measuring accuracies, but rather different cases can occur and distinctions are made between these cases.

Additional support

Map matching or smartphone sensors, such as the orientation sensor, and the GPS can provide additional support.

The accuracy of the locating procedure can be increased through map matching. To do this, a measured position is compared with a digital, georeferenced map, and the most likely position of the object on the map is determined. This procedure is particularly useful for routing from an actual position to a target position, because the position determined on the map is crucial for navigation.

The smartphone's orientation sensor can also increase localization accuracy. In most parking garages the routes have right angles, so that a 90° direction change can mean that the driver has turned. A low-pass filter is used to reduce the random noise of the sensors.

Sensor amalgamation

A particle filter is used for sensor amalgamation. In this regard, in the initial status the particles are first distributed randomly around a start position. As a rule, this is located at the entry of the parking garage or in front of the entry barrier. Then the particles are moved in a direction that depends on the actual orientation and the speed of the vehicle.

Position determination takes place thereafter, and the particles get an appropriate weighting that is relevant for stochastic universal sampling. For stochastic universal sampling all particles are considered as an individual and get a section on a line that corresponds to their weighting. Then a specific number of pointers are distributed uniformly along the line. The number of pointers equals the number of particles that are selected. The first pointer is located on a pulled random number, all other pointers are located at the same distance from each other after the first pointer.

Through stochastic universal sampling a specific number of particles is selected and new particles are generated in accordance with the weighting. The actual position equals on the particle with the greatest probability.

After determining the position, map matching takes place, in which the determined position is compared with the digital map and placed on the appropriate path.

5.2 Account

Another focus of trive.park is the user account, with which the user can personalize the application. Through the account, in a few steps it is possible for the user to make adaptations so that recommended parking spaces meet the user's requirements. Special authorizations are also associated with the account. Thus in the account it can be specified that a user has the authorization to open the barrier of a parking garage as shown in Fig. 6.

Fig. 6: Selection of the parking space and opening the barrier in the field trial

The user can specify extras for personalization of the parking areas. For example, the driver of an electric car can specify that he is looking for free parking areas with charging columns. In this case the app will only recommend such parking spaces to the user.

To increase the customer benefit, with the aid of sensors, the actual occupancy of parking areas is tracked and only free parking areas will be recommended to the user. Moreover, in a database, all parking spaces must be specified with additional information. As soon as the user enters a parking garage, he receives the suitable recommendations and is navigated directly to a free parking space. Since it is possible that there is inadequate Internet connection in the indoor area, the user receives multiple parking space recommendations. As soon as the user drives past a recommended parking area, he will be navigated to the next free parking area.

6 Outlook

After initial field trials in 2015, in the 1st quarter of 2016, a pilot project started in Wolfsburg in a semi-public parking garage shown in Fig. 7. In this project an adaptation of the trive.me, "trive.park" service was adapted for a company. The parking garage has more than 1400 parking spaces on 4 levels and 10 charging columns for E-vehicles. The objective is to develop a parking system for employees and customers, and thus implement an intelligent parking process through networking between vehicle, user, and infrastructure. What is not considered in this project is the payment process, as parking is free of charge in this facility.

Thanks to real-time monitoring of the individual parking spaces, a direct routing to the free parking space can be executed. In a subsequent expansion phase, even navigation to the destination of the adjacent company is planned for the project. The parking space is selected based on the user account, which includes the vehicle dimensions and the requirements of the users.

Fig. 7: Pilot project – Parking Garage in Warmenau/Wolfsburg

In the first step the service is integrated directly on the smartphone for iOS and Android (e.g. activity button for barrier opening). Navigation then occurs by voice command that is sent by Bluetooth to the vehicle and read aloud (similar to Bluetooth Convenience Telephony).

In a subsequent expansion phase a prediction model will be designed that can predict the probability of free parking spaces. Through precise real-time monitoring there is sufficient data to meaningfully validate this model and derive an accuracy statement.

Furthermore, as a modern infrastructure the intelligent parking garage is suitable as a test environment for automated parking of vehicles in conjunction with the infrastructure communication.

Then scaling to larger regions is planned through a minimal solution that can be extended through local adaptations.

By mapping the described use case, trive.park is a driving force within the digital transformation described above. On one hand the parking garage that is becoming more intelligent becomes part of the city in transformation, and thus in the medium-term enables highly automated parking. On the other hand, the vehicle itself is transformed into the digital world through the use of added value services that occur through networking.

Impact of demonstrated remote attacks on security of connected vehicles

Markus Ihle, Bosch Center of Competence Security, ETAS GmbH

Benjamin Glas, Bosch Center of Competence Security, ETAS GmbH

© Springer Fachmedien Wiesbaden GmbH, ein Teil von Springer Nature 2018
R. Isermann (Hrsg.), *Fahrerassistenzsysteme 2016*, Proceedings,
https://doi.org/10.1007/978-3-658-21444-9_8

Recent Remote Attack

The recent [3] successful attack on a current connected car has proven the practicability of remote attacks on connected vehicles. It is no longer a theory that can be dismissed or ignored in the development of automotive systems. The effort involved was not beyond what is to be expected for some attackers with a strong motivation.

Notably, this was not the first demonstration on a connected car. In 2010 a team consisting of security researchers of the University of Washington, Seattle and the University of California, San Diego, showed that having access to vehicle-internal communication systems gives influence on or even limited control over safety-relevant vehicle functions like longitudinal (acceleration and braking) and lateral (steering) acceleration of the car [1]. Based on this work, the same group in 2011 presented various remote real-world attacks on a "moderately prized 2009 model sedan" [2], achieving access to the internal communication systems e.g. via playing manipulated mp3-Files on the CD-player of the car, or injecting malicious code via an externally triggered audio phone connection to the car.

However, these revelations mostly stayed in the scientific community and received only limited public attention, partly due to the lack of technical details in the publications. This changed by 2015 at the latest, when Charlie Miller and Chris Valasek presented a remote attack on a 2014 model SUV in much more detail [3], and a broad media coverage brought the topic to public awareness. Here again, the authors attacked a remote connectivity feature to ultimately get full write access to a safety-relevant internal communication bus (in this case the chassis CAN bus) and thereby access to the functional interfaces using this bus. Since the attack is somehow prototypical, we look at its structure in some detail to give background and motivation for the approaches and countermeasures covered in the remainder of this contribution. For more details we refer to the original publication [3]. Figure 1 depicts an attack overview along the E/E-architecture of the affected vehicle.

The attack consists of multiple steps and took the researchers several months of intensive work to discover vulnerabilities and exploit them. At first, an unprotected application interface in the Head Unit offered access via both WLAN and cellular network and allowed execution of arbitrary code. Exploiting this vulnerability gave the attacker full control over the infotainment system including the direct interfaces like infotainment display and interior audio. As can be seen in figure 1, the infotainment unit also has direct connection to both internal CAN buses present in the vehicle. But, internal communication is done by a separate microcontroller allowing only a defined set of messages sent and received by the head unit. So as a next step, the attackers found a way to reprogram the microcontroller to overcome this restriction and gain the ability to send arbitrary messages. This leveraged an existing reprogramming interface and a lack of authentication of the flash image. Having achieved communication access the publication showed just examples of possible influence on onboard system behavior by using standard functional and diagnosis interfaces.

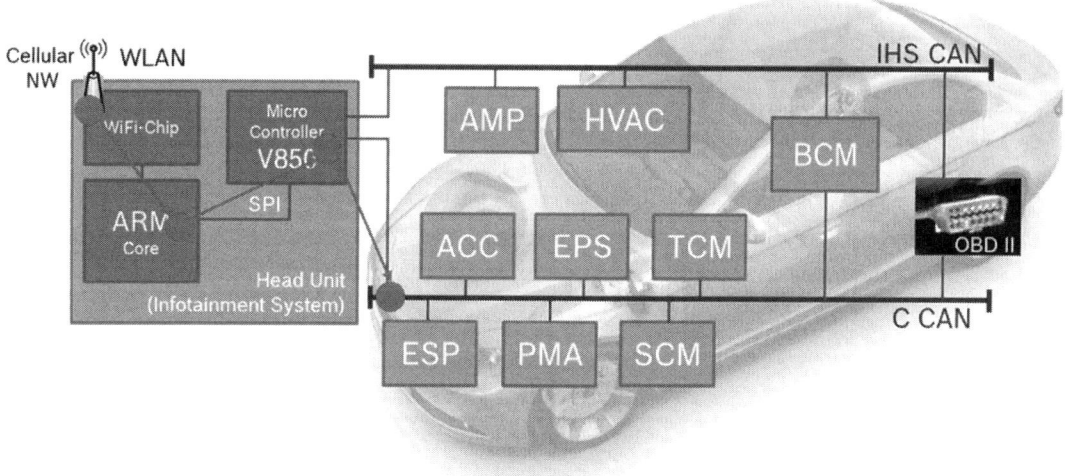

Figure 1: Path of attack by Miller and Valasek

Impact on traditional Security Assumptions

One of the cornerstones in automotive security threat and risk analyses has been the assumption that a successful remote attack is "not possible" or "too unlikely to be considered".

This could be claimed true up until about 10 years ago, as cars were not "connected" to the outside world while in normal operation. Connectivity was restricted to visits to the service station. Thus, the assumption "internal networks are secure and trustworthy" was created and valid.

With the introduction of the first Internet connected systems, the trustworthiness of the internal communication was still seen as valid if the connectivity unit – typically the telematics or head unit – introduced firewall-like functions that targeted to filter incoming IP-based communication to the car. So for many engineers, the assumption about trustworthy and secure internal communication still held true.

As almost all car architectures today get connected with the Internet, either by the manufacturer's connectivity solutions or aftermarket solutions, non-connectivity is not a valid argument anymore. The expected use has changed towards a connected architecture.

In addition, the recent attack of Miller and Valasek clearly demonstrated – again – that even though there was the attempt to filter incoming communication, the approach did not guarantee an authentic internal communication. The separation function intended to guarantee protection was itself not thoroughly protected.

From the authors' perspective, former paradigms like "internal networks are secure and trustworthy" do no longer hold and have to be questioned and re-evaluated in every future risk and threat analysis.

From Single Line of Defense to Defense in Depth

A single line of defense is dangerous in security. Even if we assume that the theoretical design of a security control is flawless, the implementation rarely ever will be. Implementations are prone to implementation errors or "bugs". It is empirically known that bugs are one of the main drivers for vulnerabilities, both in security relevant and security non-relevant parts of a system. Besides academic approaches based on very limited functionality, there is yet no such thing as a perfectly implemented complex system. A lesson learned in the software industry decades ago.

If we accept the fact that every countermeasure – or every line of defense – itself is likely to have vulnerabilities based on implementation errors, ensuring the security of a system by just this single line is obviously not an approach that promises high robustness.

What we need is a defense in depth!

Like many medieval castles and the fortresses built by the French master builder Vauban in the seventeenth century, we need protection mechanisms in a certain depth to still control an attack even if the first wall has been breached.

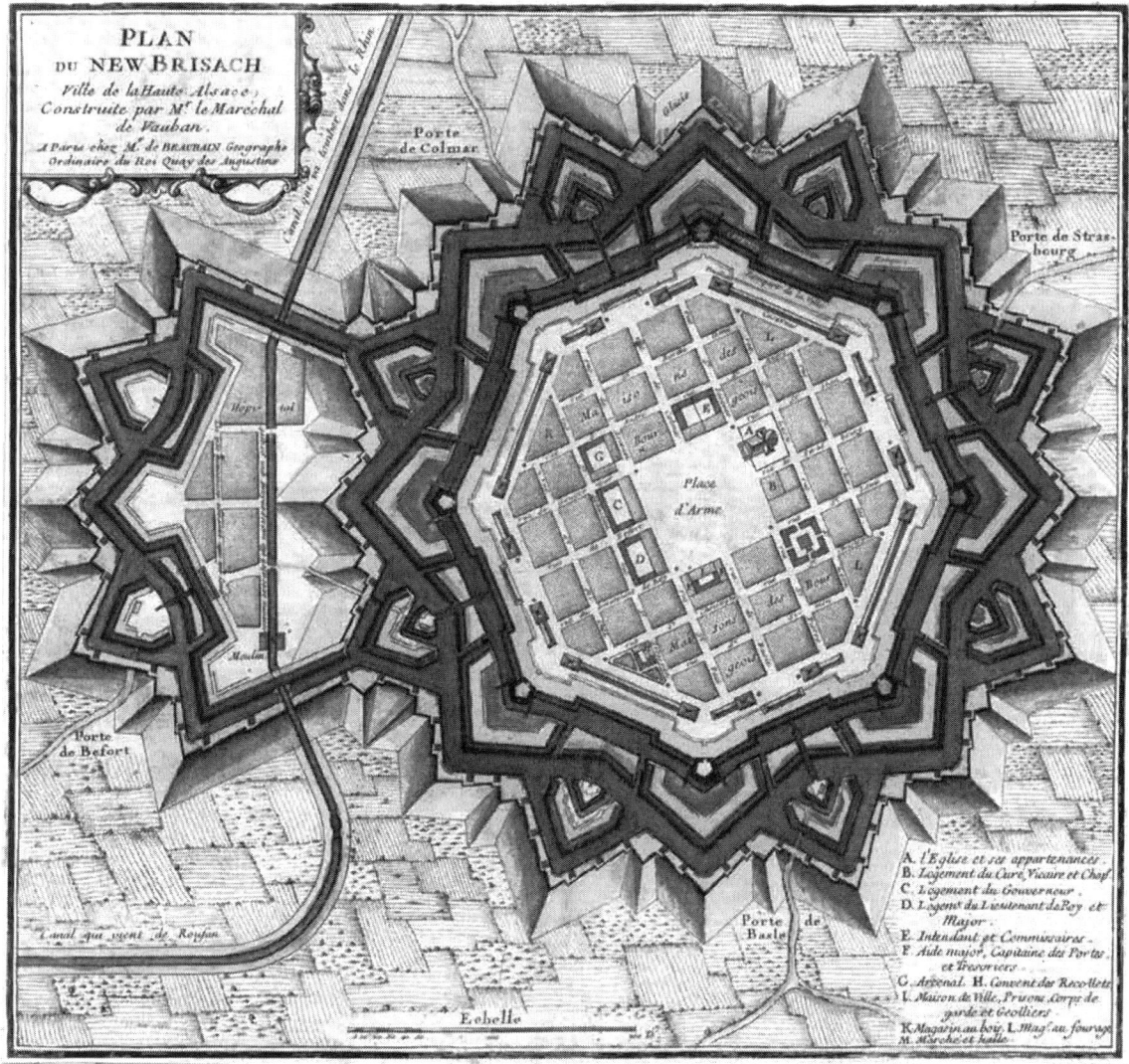

Figure 2: A fortress built by Sébastien Le Prestre de Vauban, ca. 1698 [4, 5]

In the IT industry firewalls, network zones, separation, etc. are usually used in parallel to be able to minimize the impact on the attacked system – even if some security mechanisms have been penetrated by the attacker.

Also, the likelihood to successfully overcome multiple protection mechanisms within the existing constraints like time, resources and available competencies is lower than it is to find just one weakness in a single implementation of one protection mechanism.

Security to protect Safety

A lot of the systems in a car are safety-relevant: longitudinal and lateral acceleration (accelerating, braking, steering), active (emergency braking) and passive (airbags) safety systems being the most obvious. But even comfort-oriented features can impact driver safety: external light, internal light, speaker volume, and wipers can at least distract and cause undesirable consequences. From the authors' experience the vast majority of the electronic control modules (ECUs) contain such functionality.

The protection of the intended functionality of these safety-relevant functions inside ECUs is an important property: Only a system with a well-defined and valid internal state can be rated as safe. Safety requires the integrity of the system, both with respect to random failure as well as against expected manipulation [10]. Integrity protection of ECUs is a typical security design goal and an important base requirement for the overall security architecture.

Since the introduction of car-internal networks, safety-relevant functions get interconnected, for example the engine's torque is reduced during braking. Some functions are even completely built upon interconnection – as it is with adaptive cruise control (ACC), where the ACC directly controls engine control and braking via network commands.

Increasing automation will strengthen this trend, having centralized powerful ECUs controlling the behavior of multiple other ECUs – and in the end directly controlling the vehicle's trajectory.

The car-internal networks like CAN, FlexRay or Ethernet are carrying the information between these safety-relevant ECUs. The communication link often inherits some safety-relevance from the system requirements of the ECUs and functions they connect.

It is therefore a much more important design goal to protect the integrity and authenticity of the ECUs and their communication against deliberate manipulation as it was before the introduction of connectivity.

The Bosch Four-Layer Model to vehicle security

The system to be secured in Automotive Security comprises the vehicle itself and its connected environment, including backend, attached CE-devices, and more. In addition, every sound security concept for a complex system like a car needs to be holistic, considering the overall vehicle architecture, the complex supply chain, and the long lifecycle including all the different phases. Nevertheless, for this contribution we are going to focus on the vehicle security architecture.

From an architecture point of view, Bosch is suggesting a multi-layered approach to car security. While the vehicle consists of a plethora of parts, the security of the vehicle is more than the security of its components. It depends on a consistent security architecture and a tailored and reliable interplay of security mechanisms and components. There is no single "silver bullet" solution for security architectures, but multiple ways to achieve a similar level of security by using different choices of protections, mechanisms and implementation locations, even more so since vehicle architectures, features and properties differ between carmakers, models and equipment variants. There are basic rules and patterns that can be used for building secure systems, but in fact each security concept and architecture has to be tailored to the specific system to provide a both sufficient and economically feasible protection.

Hence a close cooperation and communication between system responsible – usually the vehicle manufacturer – and suppliers is necessary to make sure that each component is aware of both the security requirements the concept poses on it and the correct security assumptions it can make on its environment. Only then the subsystems and components can be built to reliably perform their part in the overall vehicle security concept.

The vehicle manufacturer as system responsible would typically develop and define the overall security concept and derive the matching security requirements and boundary conditions for the individual components. Tier 1's responsibility would be to match the information with his own analysis and to ensure a match of his concepts and requirements with those imposed by his customer. Bilateral exchange should ensure that the whole security architecture is correct and fulfils the security goals of the car security concept.

Assuming this is all in place, we still know that every component can have weaknesses. In software engineering, estimations of security weaknesses vary between one per 1,000 and one per 10,000 Lines of Code (LOC). Considering that a current upper-class vehicle may contain well over 100 million LOC, the existence of errors and weaknesses can be seen as a fact. Therefore, a defense-in-depth concept is crucial. This way, an attacker can be forced to find several vulnerabilities in different systems before achieving his or her attack target. In addition, the effect of (partly) successful attacks can be limited and damage contained.

Our layered approach achieves both – combining different components to an overall security concept and implementing a defense-in-depth strategy. In the following, we present four major layers in a general vehicle security concept and give some examples of major objectives and possible mechanisms on the respective layers. Figure 3 depicts an abstract overview.

7

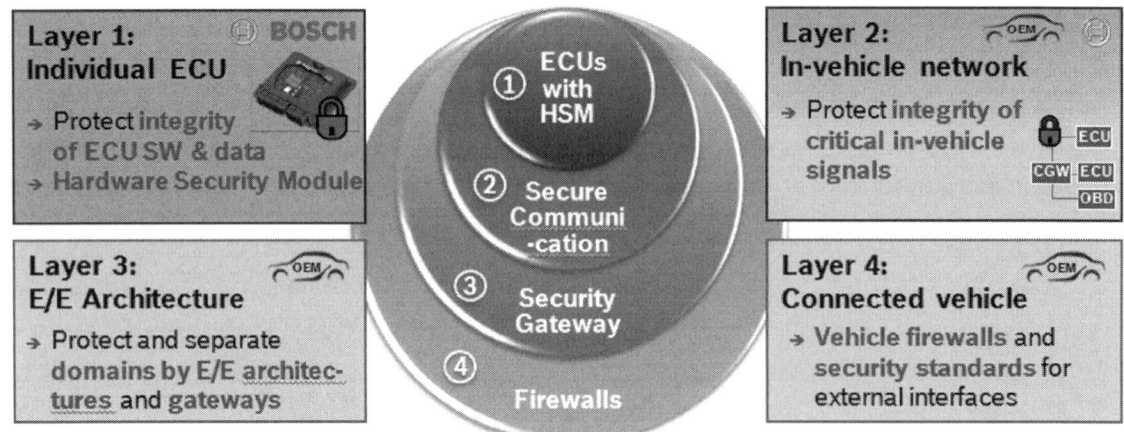

Figure 3: Overview of Bosch Multi Layered Concept

The different layers have to complement and support each other to create a consistent security architecture. So security controls have to be selected and put in place to fulfill security functional requirements. On top of that, the security functionality has to be protected itself, since attacking the protections is a straightforward and to be expected attack strategy. Therefore especially the protection on layer one fulfills both a direct and indirect protection property as shown in figure 4.

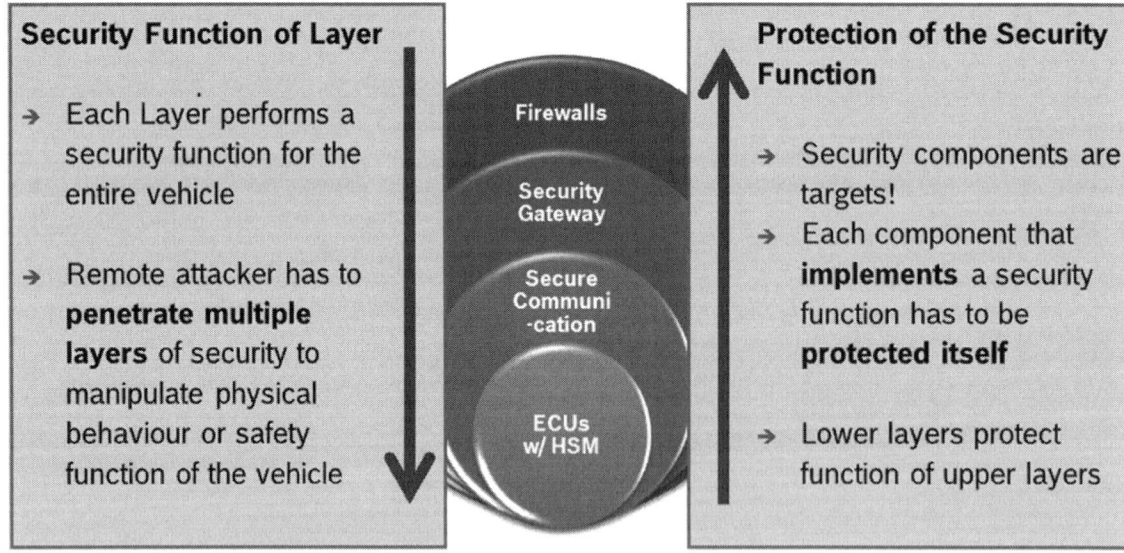

Figure 4: The mutual impact of the Security Layers

Layer 4 – Protect the Vehicle as a node in a global network

With rapidly increasing connectivity features, vehicles have to be assumed connected devices in a global network rather than isolated, closed systems. While offering huge opportunities for new features and functionalities, this remote connectivity tremendously enlarges the attack surface of the vehicle and makes vehicles susceptible to remote attacks. This is a whole new dimension since the former attacks that mostly required physical access to the vehicle don't scale the way a remote attack does. Remotely, a single attacker could potentially attack entire fleets of vehicles instead of only single instances.

Consequently, security has to respond to this threat and protect the vehicle as a node in the remote network. Since this problem is well known in the area of classical computer networks and the Internet, protection mechanisms are readily available and can be reused from this domain.

Primary component in this layer is the device that directly links the car to external networks, e.g. a Telematics Control Unit (TCU) or an infotainment system like a Head Unit (HU). Also local connection point to connected devices like smartphones have to be considered here. Protecting these external communication points creates a kind of perimeter security for the vehicle.

One main challenge is that two domains touch in the infotainment system. One is the classical "automotive" domain, which is real-time centered, statically configured, and runs safety-relevant control systems with direct influence on the main vehicle functions. The other, let's call it the "connectivity" domain is more Internet and CE-device centered, allowing remote connectivity, communication with CE-devices like smartphones, and perhaps personalization like download and execution of Apps. Also the latter has to cope with much shorter development cycles and more rapid and dynamic development of functionality. In order to protect the critical automotive domain from unwanted and potentially harmful influence, we propose the domains to be strictly separated. Figure 5 depicts a possible approach, the Bosch "Dual Architecture". This can be realized in hardware or software, e.g. using a hypervisor for effective separation.

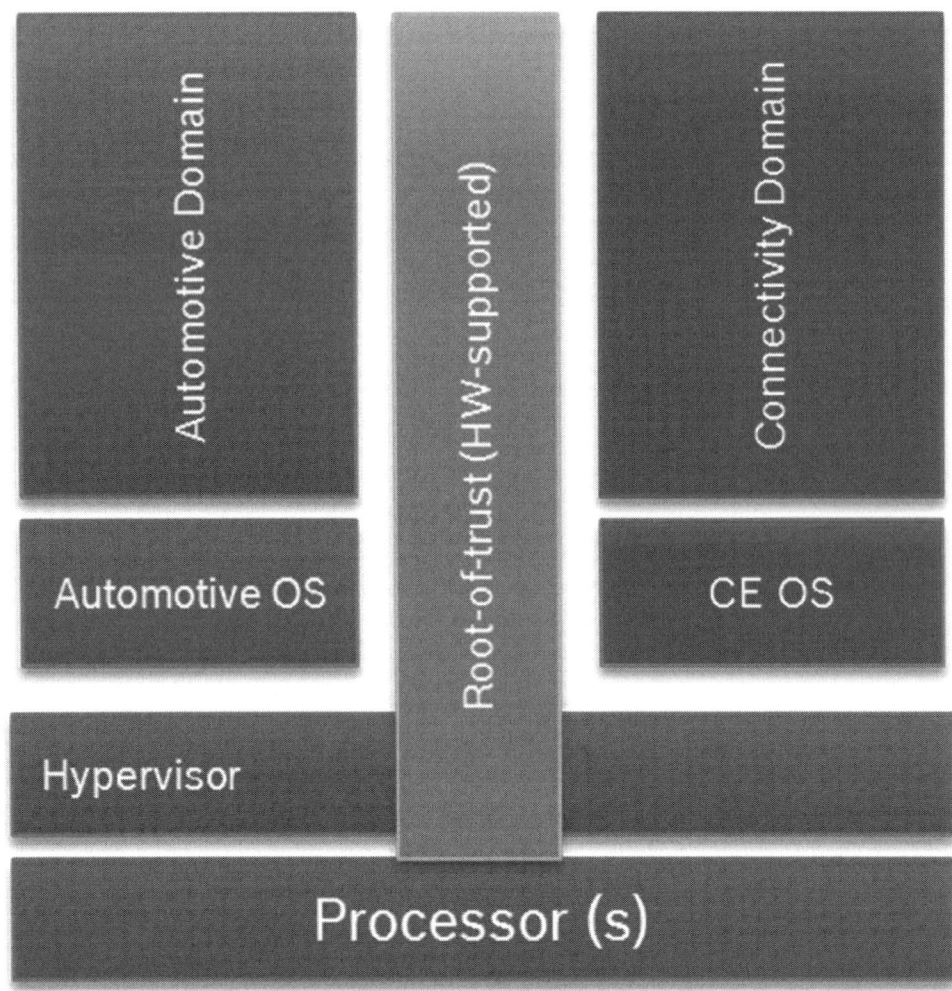

Figure 5: Bosch "Dual Architecture"

Communication between the domains should be strictly controlled, using for example classical firewalling approaches including filtering of traffic.

Also an application layer gateway can be used to terminate any inbound traffic and restrict internal communication to clearly defined connections and protocols.

Layer 3 – Leverage the internal E/E Architecture

A crucial aspect both for building a barrier against external influence to internal systems and to restrict the impact of compromised subsystems is using and protecting the internal electric/electronic communication architecture. Central component can be a so-called "central gateway", usually a unit connecting the major internal communication systems like different CAN busses (e.g. powertrain, chassis and cabin CAN), FlexRay busses and ethernet. Also all external communication ports as TCUs and diagnosis (e.g. OBD II) interfaces should be separated from the internal communication by a gateway. Again separation based on classical firewall approaches offer protection for critical internal systems by effectively restricting communication to the intended functional traffic.

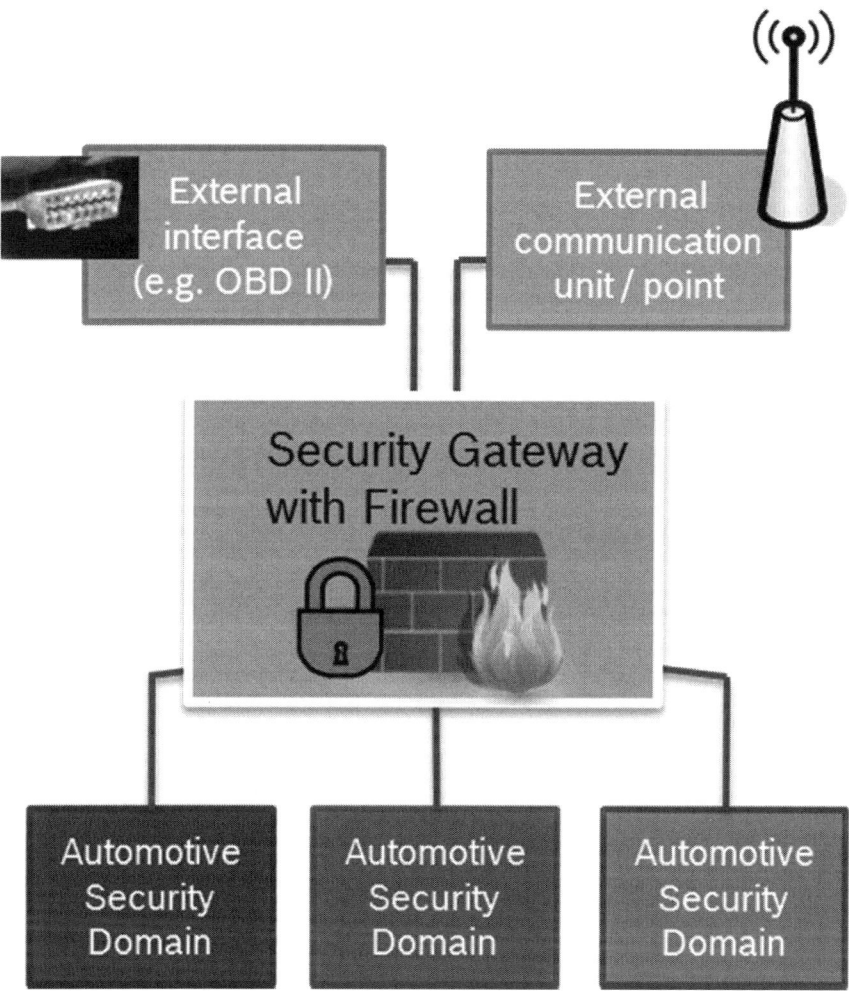

Figure 6: Domain separation with a Security Gateway

Additional communication that is needed only in special situations like diagnosis or reprogramming should be possible only after proper authentication and authorization and if needed verification of proper environmental situation. Here various parameters could be used, e.g. maximum vehicle speed, state of certain systems, presence of certain tokens (keyfob, smartcards…), a.s.o.

An additional element that is more and more prominently discussed recently is the use of intrusion detection systems (IDS). While IDS could be implemented on any of the layers, mostly monitoring of the communication is discussed as it is also common in business networks. In the automotive context this approach is relatively new and still some development is needed to achieve systems that reliably detect attacks without

11

producing too many critical false positives. This is especially relevant since the immediate questions arises how to act upon detected attacks. Pure IDS only report detected irregularities. For this the required infrastructure both inside the car and in respective backends have to be built up. But when talking about safety-related systems under attack, of course additional immediate and automated reactions have to be considered, switching from IDS to an Intrusion Protection System (IPS). Here false positives triggering unintended reactions could be fatal. Therefore, still some research has to be invested for these systems to become fit for real-world application.

Layer 2 – Secure the vehicle-internal communication

Assuming that an attacker is able to get access to internal communication systems, for example by compromising and taking control of any of the connected ECUs or obtaining access via a Wifi-dongle plugged into the OBD II port, he is able to participate in the respective communication, which may well be safety-relevant. Depending on the compromised component and the used communication system he may be able to use functional communication and interfaces to:

- monitor communication, e.g. for person related data (breach of privacy)
- add communication, e.g. transmit faked brake commands to the respective ECU
- manipulate information between sender and receiver, e.g. add an offset to an engine torque signal

Based on the common assumption about separated internal communication, up to now most internal messages are only protected against random errors, not against deliberate manipulation by an attacker. So without additional countermeasures, it is easy for an attacker to misuse the existing intended functionality and to create a safety impact.

In order to meet these threats, a number of well-established security controls and mechanisms can be used, including cryptographic protection of messages, depending on the protection goal. In many cases, especially for safety-related messages, authenticity of sender and data is desirable. To achieve this, cryptographic message authentication codes (MACs) with proper key hierarchies and message counters can be used to prove authenticity of the sender, and integrity and freshness of the data transmitted. Technical approaches for this kind of secure on-board communication have been standardized and are described since AUTOSAR 4.2.1. [6].

If content shall be protected against monitoring, e.g. to protect privacy-relevant data, confidentiality can be enforced using encryption. Proper and established primitives and protocols are available but have to be chosen carefully according to security needs and automotive constraints and requirements.

Layer 1 – Protect the individual ECU

Protection of the individual ECUs is the foundation of the layered security concept, achieving two main goals: Protection of the integrity and intended functionality of the device, preventing misuse, and protection of the security functionality implemented on the device. Latter is especially important for crucial security components like "perimeter devices" (e.g. TCU or HU) and central gateways.

Multiple elements and controls can be used to together ensure integrity of devices as the primary goal. Interfaces for debugging, diagnosis and reprogramming are needed, but have to be protected against misuse. Updates should be possible but have to be protected against manipulation. And depending on the relevant attacker model the device has to withstand attacks only on the communication interfaces, direct physical semi-intrusive or intrusive attacks, and indirect attacks, for example via side channels.

Again – approaches are available and can make use of many established security primitives and controls. But, the special automotive requirements with safety requirements and harsh constraints on resource usage, real-time properties, and cost also pose challenges to security solutions.

A sensible approach could be to protect sensitive interfaces with appropriate authentication, secure flash content and update files with software signatures, and check the integrity of the system also during operation. Depending on the device both mere detection and logging of integrity violations or direct reaction can be reasonable, even down to disabling the device or restricting its functionality.

To implement security, there are several options and different building blocks are required for different levels of protection. In general, secure storage for keys (and logs) is desirable as well as a secure execution environment for the security functions and cryptography, perhaps supported by specific hardware. Multiple approaches are available on the open market. Systems using microprocessors can make use of vendor- or architecture-specific security extensions like ARMs TrustZone [12] or Intels virtualization extensions VT-x [11].

For microcontrollers, several de-facto-standards are available in many open-market devices. A minimal security peripheral offering protection for symmetric keys and hardware-implemented AES [8] called "Secure Hardware Extension" (SHE) is specified by the German "Herstellerinitiative Software" HIS. This SHE can be seen as an implementation of the "EVITA-small" module described by the EVITA-project [13] for use in smaller ECUs and sensors to protect communication.

For larger or more critical devices Bosch specified an automotive Hardware Security Module (HSM) [9], sometimes referred to as "Bosch-HSM", which is available from various vendors. Basically this comprises a dedicated programmable security core with its own internal flash and RAM together with additional security peripherals like a True Random Number Generator (TRNG) and an AES accelerator in hardware. This HSM is very similar to the "EVITA-medium" module recommended to offer enhanced protection.

Security Maintenance after start of production

Security can be seen as three generic steps:

- protection – to harden the system against attacks
- detection – to detect attempts or successful attacks
- response – to counter detected attacks

Most of what has been written before can be assigned to protection of the system, designing the system in a way that it is robust against manipulation of the intended behavior. After development and start of production (SOP), this security robustness gradually decreases over time, as the capabilities of attackers and the effort and cost involved in an attack decreases, e.g. by cheaper computing performance, the increase of technical and mathematical knowledge or by more precise measurement equipment that offers new possibilities.

One effect is the occasional "breaking" of cryptographic primitives that were rated as "secure" before by the security community, but suddenly are not state of the art anymore. An example is the "MD5" cryptographic hash algorithm that was in legitimate use only for about 10 years. Today, the long-used alternative "Secure Hash Algorithm" SHA-1 is still not completely broken, but most experts would recommend a switch to SHA-2 or SHA-3 for cryptographic hashing [7].

If we think about the complete life cycle of an ECU for a car, 10 years are way too short to be sure that the initial selection during product development is still valid towards end of life of the car. Therefore, a primitive or protocol broken after being designed into the system needs to be replaced by a sound alternative. That means a software update has to be developed and deployed to protect the car against possible attacks.

In addition to the changes in the state of the art in cryptography – and statistically much more often – existing vulnerabilities in a product will be discovered after start of production. Being introduced to the system by implementation errors, so-called "bugs". Vulnerabilities are entry points for an attack when exploited.

As it seems unlikely that complex systems consisting of hundreds of thousands or million lines of code will be bug-free in the near future, the fixing of the identified bugs and the patching of the deployed car fleet will be a necessity. Even many years after SOP.

What we need is long-time maintenance for the security of connected cars. As we are used to it from our operating systems on our PCs, a piece of software unmaintained cannot be seen as robust or reliable against attacks after a certain point in time.

Business Model Impact

The need to detect security vulnerabilities and to respond to them by the creation of updates in a timely fashion and over very long time periods will be drivers to develop new requirements and capabilities for the automotive industry as a whole.

Today many automotive products are not planned to be in an extensive maintenance phase over a time period of 10 or 15 years. Traditionally they are more "one-shot" developments. The product is developed and tested and finally produced and shipped for years with an absolute minimum of changes. The original development team moves on to develop the next product.

Thereby, software development is often a single invest, not a continuous effort. Worse, software is seen as "just some functionality in the hardware" and the hardware production often is dominating the price. If we accept that a complex piece of software needs to be maintained, the business model should adopt accordingly. In the IT industry, software is sold including maintenance contracts and maintenance fees. Having a stable stream of turnover enables to support dedicated maintenance teams and fast updates.

What we propose is relevant for multiple customer-supplier relationships: Tier 2 to Tier 1, Tier 1 to vehicle manufacturer and manufacturer to consumer – it is the same challenge on all levels. Although new and rather unexpected for the consumer, it seems a realistic and legitimate approach to bind certain critical and powerful functionality like connectivity of a vehicle with a maintenance contract.

Firmware Over The Air (FOTA) updates

We already explored on the fact that there will be a need for updates of deployed systems, most likely even frequent updates. Taking into account that the updates are needed to remove existing vulnerabilities, the time needed to get a high coverage of updated systems in the field is an essential requirement. The slower this process, the more exposure of consumers to a potential attack.

The current state of the art – an update of the car in a service station, about once a year, does not seem sufficient, as it would mean on average more than six months of exposure per vulnerability. Neither is a frequent recall a viable solution due to the cost implied and the feeling of insecurity that would be created at the consumer. Both cost and consumer insecurity should be minimized.

Secure firmware over the air will give the automotive industry all of this: A relatively cheap and fast option to remotely update security functionality. The basic precondition – connectivity – is already there.

Summary and Outlook

With the recent attacks on remote vehicles, the relevance of such attacks cannot be discounted anymore. Traditional assumptions in vehicle security can hardly be continued, like the misconception of inherent security of internal bus systems.

By introducing protection mechanisms on multiple levels and maintaining them as described after SOP, the automotive industry will be able to keep the residual risk of remote attacks at acceptable low levels during the complete life cycle of a car. The required maintenance effort has to be funded by proper business models, for example maintenance contracts.

To make this work, we need multiple parties in the supply chain to participate responsibly with their part for the security of the "system car". It is the car manufacturer who has the ultimate responsibility for the whole system, but he is building upon components provided by the supplying industry.

Today, the execution of this distributed responsibility is not as clearly standardized as it should be. The work interfaces between the parties as well as the engineering processes to design secure systems have to be described and standardized. Bosch is proposing to have a normative standard at ISO and is therefore supporting the ISO NWIP 3556 "Road Vehicles – Automotive Security Engineering".

References

[1] Koscher et al: Experimental Security Analysis of a Modern Automobile, 2010 IEEE Symposium on Security and Privacy

[2] Checkoway et al: Comprehensive Experimental Analyses of Automotive Attack Surfaces, USENIX Security 2011

[3] Miller, Valasesk: Remote Exploitation of an Unaltered Passenger Vehicle, August 10, 2015 (http://illmatics.com/Remote%20Car%20Hacking.pdf, accessed February 23, 2016)

[4] https://en.wikipedia.org/wiki/S%C3%A9bastien_Le_Prestre_de_Vauban, accessed on February 23, 2016

[5] https://en.wikipedia.org/wiki/Neuf-Brisach, accessed on February 23, 2016

[6] AUTOSAR Release 4.2 Overview and Revision History, http://www.autosar.org/fileadmin/files/releases/4-2/AUTOSAR_TR_ReleaseOverviewAndRevHistory.pdf, accessed on February 23, 2016

[7] FIPS PUB 180-4, Secure Hash Standard (SHS), http://nvlpubs.nist.gov/nist-pubs/FIPS/NIST.FIPS.180-4.pdf, accessed on February 23, 2016

[8] FIPS PUB 197, ADVANCED ENCRYPTION STANDARD (AES), http://csrc.nist.gov/publications/fips/fips197/fips-197.pdf, accessed February 23, 2016

[9] Bubeck et al: A Hardware Security Module for Engine Control Units, ESCAR Conference, Dresden, Germany, 2011

[10] Kriso, Ihle: Automotive Security im Kontext der Funktionssicherheit (ISO 26262), 31. VDI/VW Gemeinschaftstagung Automotive Security, Wolfsburg, 2015

[11] Intel® 64 and IA-32 Architectures Developer's Manual, http://www.intel.com/content/www/us/en/architecture-and-technology/64-ia-32-architectures-software-developer-manual-325462.html, accessed February 23, 2016

[12] ARM TrustZone, http://www.arm.com/products/processors/technologies/trustzone/, accessed February 23, 2016

[13] EVITA, E-safety vehicle intrusion protected applications, http://www.evita-project.org/, 2008

17

Auto update –
safe and secure over-the-air (SOTA) software update for advanced driving assistance systems

Markus Koegel, Marko Wolf
ESCRYPT GmbH – Embedded Security

© Springer Fachmedien Wiesbaden GmbH, ein Teil von Springer Nature 2018
R. Isermann (Hrsg.), *Fahrerassistenzsysteme 2016*, Proceedings,
https://doi.org/10.1007/978-3-658-21444-9_9

1 Auto-Update for ADAS

From the outside, today's cars do not differ too much from 1990's models. Under the hood, however, a real revolution has taken place – not just with respect to safety and environmental aspects. Software-based *Electronic Control Units (ECUs)* in today's vehicles have taken the places of nearly all mechanical steering components and made a huge leap in convenience and cost efficiency possible. ECUs are interconnected by one or more networks and share information that enable complex interactions and regulations. This covers also, for instance, potentially safety-critical *Advanced Driver Assistance Systems (ADAS)*, such as the *Adaptive Cruise Control (ACC)*, automated parking or lane departure warning. The increasing functional capabilities, however, come along with a higher system complexity, which makes software bugs more likely to occur that might have implications for the safety and security of the whole vehicle. However, with regular software updates and enhancements one can ensure the systems' correctness, efficiency, and reliability over the complete lifetime of the vehicle. Regular updates, on the other hand, lead to additional garage visits or, in serious cases, to vehicle callbacks. In the best case, this is merely inconvenient for the customer, but in any case, it is time consuming and expensive. Moreover, these on-site updates often come too slow – especially in case of potentially safety-critical ADAS malfunctions. Also, such updates seldom reach all affected vehicles in time, so that possible hazards and vulnerabilities due to the above-mentioned software bugs might remain a safety or security risk for a long period of time.

The upward trend away from isolated, stand-alone vehicles to so-called *connected cars*, i.e., vehicles with access to the Internet, could provide a solution to this challenge. Connected cars either have a direct cellular networking interface to the Internet or have a built-in Wi-Fi or Bluetooth functionality to make use of external mobile devices via tethering. The number of connected cars within recent forecasts is still only a small fraction of all the Internet-capable embedded devices. Nevertheless, the trend to connectivity needs to be regarded also for the automotive domain. This becomes clear, as several technologies are waiting in the wings to be introduced in the coming years. The *eCall* system, for instance, will be mandatory for all newly produced vehicles in the European Union from 2018 on, while the first aspects of the *C2X* technology have already been realized by connected cars.

By means of mobile communication interfaces, which enable connected cars to be Internet-capable, necessary software updates can be performed wirelessly *over-the-air (OTA)*. These OTA software or firmware updates are advantageous for both the *Original Equipment Manufacturer (OEM)* and the vehicle owner, because the vehicle can receive the service at its current location. This is, it can be updated conveniently, instantly, comprehensively, cost-efficiently, and without visiting a garage. However, adding new communication interfaces might also set the stage for new hazards or vulnerabilities. In order to prevent that this helpful OTA update functionality itself becomes a new risk factor for vehicle safety or vehicle security, it is necessary to foresee and implement effective, automotive-specific protection measures from the very beginning.

In this paper, we first describe the current state of OTA software updates for cars, important benefits, as well as the accompanying risks with regard to ADAS, while focusing especially on safety and security. In this course, we identify safety and security assets and threats and derive corresponding safety and security requirements for an OTA system. As our main contributions, we then present a description for a *safe and secure over-the-air (SOTA)* software update process, the design parameters, and a general SOTA software update system architecture, which meets the identified safety and security requirements. We conclude this paper with a summary and a short look into the future of automotive SOTA software updates.

2 An Overview of Over-The-Air Software Updates

In this section, we first review the current state-of-the-art of OTA software updates in the vehicular domain and sketch the benefits of the technology. We describe how OTA software updates work in detail and look deeper into the trade-off between update transfer efforts and the complexity of in-vehicle update installation mechanisms.

2.1 State-of-the-Art Vehicular OTA Software Update Solutions

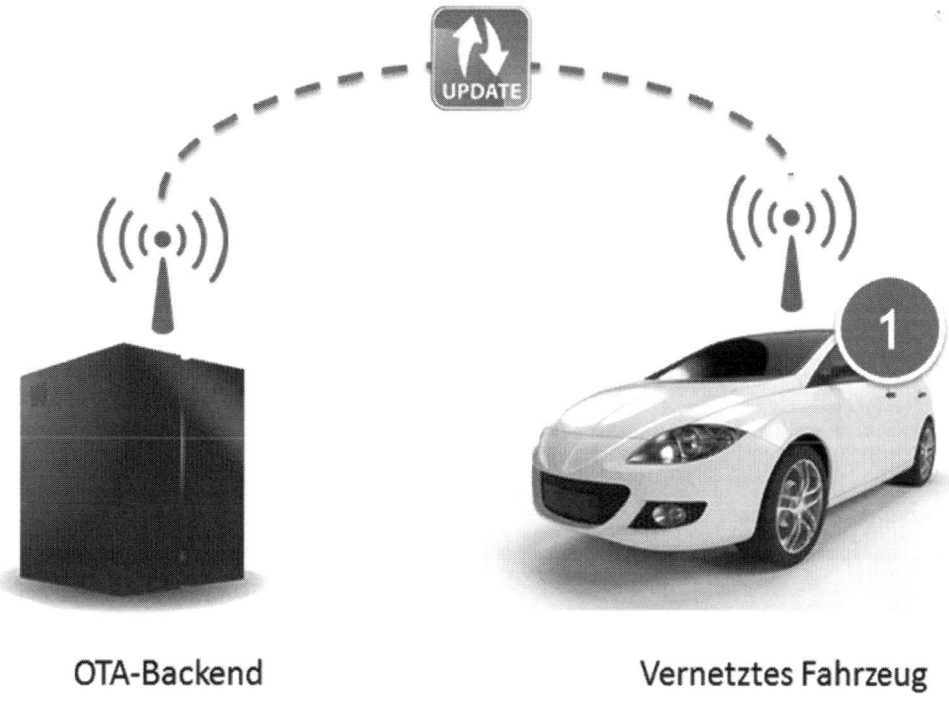

OTA-Backend   Vernetztes Fahrzeug

Figure 1: Over-the-air software updates for modern connected vehicles.

The use of OTA software updates – schematically depicted in Figure 1 – is far from being a fundamentally new technology. Quite the contrary, several standards have been proposed, for example by the Open Mobile Alliance (OMA). In practice, firmware over-the-air (FOTA) has already been utilized for years to update smart devices, such as smartphones and tablets. In 2014, BMW announced that the next generation of "Connected Drive" would feature OTA updates for the integrated navigation charts and maps [1]. Also, the search for gas stations with low prices was also introduced via OTA updates in the same year. Already in 2012, Tesla went one step further and not only updated map material, but enabled and changed even safety-critical vehicle features and parameters [2]. In January 2015, BMW also used an OTA update to fix a security vulnerability that could have been exploited to unlock the doors of up to 2.2 million vehicles [3].

2.2 Vehicular OTA Software Update Benefits

From the above description of the current state-of-the-art, the benefits of software OTA updates become clear. A manufacturer is able to keep the software also on already shipped and delivered electronic units up-to-date, without causing virtually any inconvenience for the customer. Of course, this does not only include regular functional updates, but also hot-fixes for safety-critical or security-critical vulnerabilities. With OTA updates, the customer does not need to visit a garage nor does a service technician need to visit the customer on-site. Instead, a customer connects either to a local Wi-Fi or to a mobile broadband network and directly downloads the necessary update package. As in the above-mentioned cases, the manufacturer can install *Subscriber Identity Modules (SIMs)* for the latter case, so a customer does not need to have access to a local Wi-Fi and, in consequence, a high degree of coverage can be achieved for the software update.

In addition to keeping software up-to-date, the OTA technique allows also for a range of new business models. A feature activation module, for example, is only one possible application that could be implemented: a customer could unlock another region in the navigation charts, activate a live traffic information service, or any other optional on-demand function or service. For rental services, OTA software updates for the use with feature activation offers even more possibilities. For example, a customer could remotely enlarge the area, in which the vehicle is allowed to be taken during the rental period or the engine performance could be increased temporarily during a trip over hilly ground. Aside from ADAS, the usage degree of the tools that are mounted on or being included in utility vehicles could also be controlled via OTA update-driven feature activation.

2.3 How Vehicular OTA Software Updates Work

As shown in Figure 2, an over-the-air software update process typically works as follows.

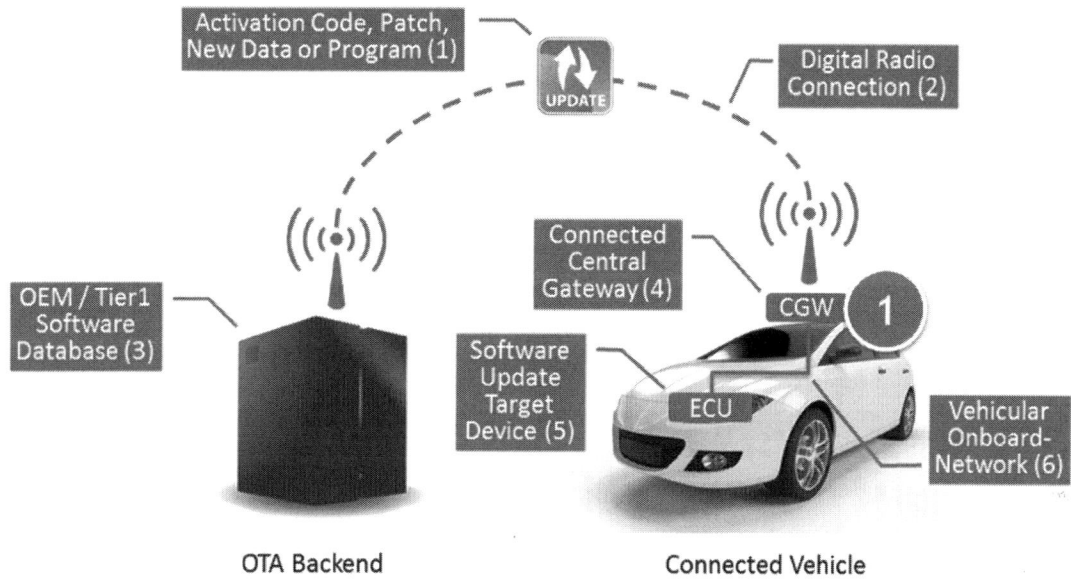

Figure 2: Vehicular over-the-air update components.

Using its wireless data connection (2), realized for instance as an integrated cellular or Wi-Fi interface at the *central gateway (CGW, 4)*, the connected vehicle regularly communicates with the central OTA backend server of its manufacturer. The vehicle first transfers its individual *vehicle identification number (VIN)*, together with its current software configuration to enable the OTA backend to check whether the current vehicle software configuration is still up-to-date or if any new software could be installed from its central software database (3). The central OTA software database might include not only OEM software, but also approved supplier software or even approved third party software. In case a new software patch, activation code, ECU program or ECU data update (1) has been identified, the OTA backend transfers this software update directly to the vehicle using the wireless link (1) independent of the actual location of the vehicle.

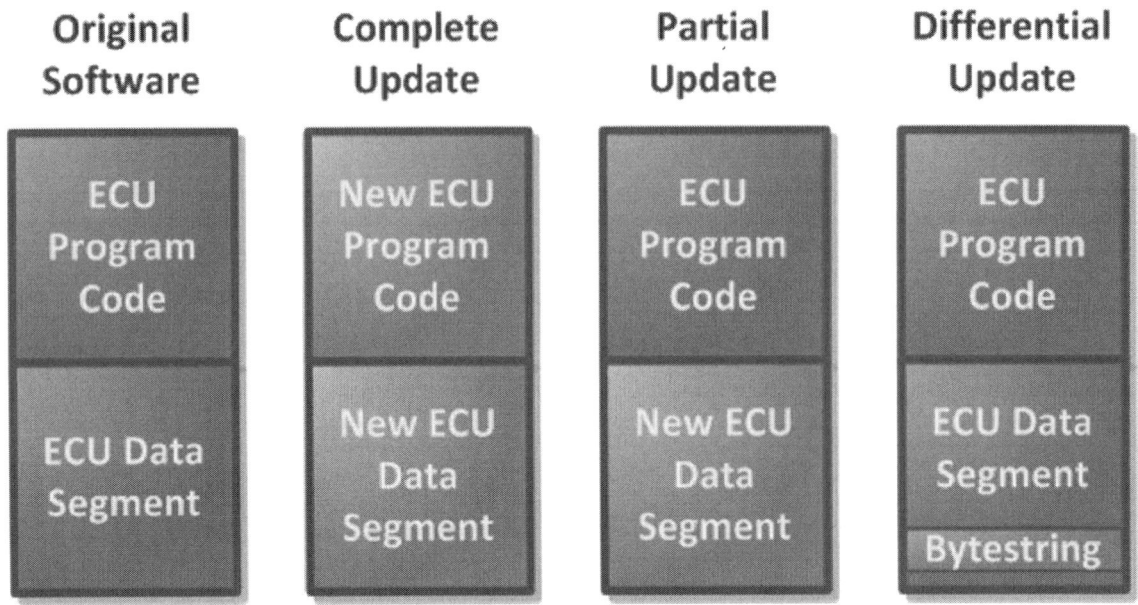

Figure 3: Complete, partial, and differential software updates.

As shown in Figure 3, such a software update for a vehicular electronic control unit (ECU) can replace:

- one or more original software applications completely (complete update) or
- only the affected portions of the original software (partial update) or
- only the affected bits and bytes of the original software based on an adequate delta update mechanism available at the backend and at the vehicle (differential update).

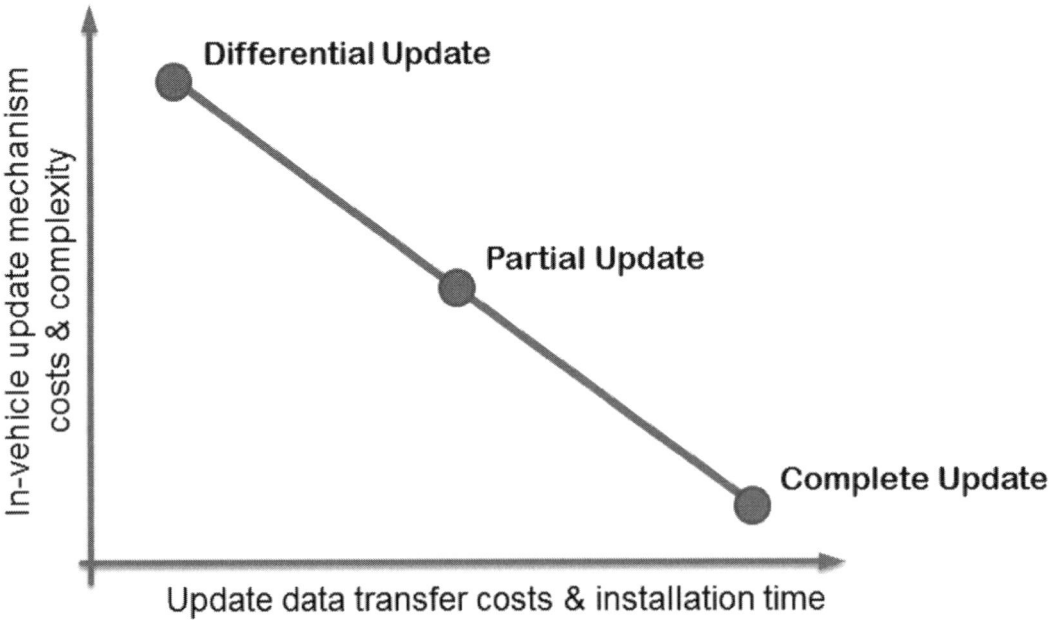

Figure 4: Systemic trade-offs of in-vehicle update installation mechanism.

There is a typical trade-off between reducing the time & amount of data needed to transfer the software update against the complexity and costs required for the update mechanism applying the software update at the vehicle ECU (cf. Figure 4). Once the update has been received completely, the vehicle checks whether itself is in a safe operational state (e.g. parking mode, sufficient battery power etc.) and then requests the vehicle driver's approval for actually installing the software update. The update process is usually executed after the driver has powered down the vehicle and takes usually between 15 and 45 minutes. As exemplary calculated at Table 1, much of the time is typically already required for transferring the update data from the gateway unit towards the software update target ECU (5), especially if the target ECU is connected only by a low-bandwidth on-board network system (6). Hence, only for highly safety- or security-critical updates, it might be possible to execute an update process also at vehicle start, however strongly limited to a maximum of 1-5 minutes. After the application of the software update, the result (i.e., success or failure) is logged internally, reported to the driver, and optionally also reported to the OTA backend.

Table 1: Average update data transfer times on typical automotive bus systems.

Onboard network	Typical gross bus speed [4]	Typical net through-put [4]	Exemplary times (transmission phase) for transferring typical update items for an ECU such as a new:				
			Patch	*Config*	*Program*	*Data*	*Map*
			1 kB	*50 kB*	*1 MB*	*50 MB*	*1 GB*
LIN	20 kbit/s	1 kB/s	1 sec	1 min	17 min	14 hours	12 days
CAN	500 kbit/s	20 kB/s	50 ms	3 sec	1 min	42 min	14 hours
FlexRay	10 Mbit/s	50 kB/s	20 ms	1 sec	20 sec	17 min	6 hours
CAN FD	4 Mbit/s	80 kB/s	<1 ms	625 ms	13 sec	11 min	4 hours
MOST50	50 Mbit/s	100 kB/s	<1 ms	500 ms	10 sec	8 min	3 hours
Ethernet	100 Mbit/s	150 kB/s	<1 ms	300 ms	7 sec	6 min	2 hours

3 Safety and Security Risks for Vehicular OTA Software Updates

While introducing a wide range of benefits, equipping a vehicle with OTA software update functionality might introduce also some new risk to vehicle's reliability and safety. It might, for instance, enlarge the so-called attack surface. This means that it might increase the possible starting points that could be exploited by an attacker.

In this section, we first highlight the implications of OTA software updates on vehicular safety aspects and then turn to the security perspective. We identify the security assets and relevant security objectives and describe the corresponding threats against these objectives. As our last contribution in this section, we postulate matching security requirements. By fulfilling these requirements, the described OTA update system can be hardened against the identified threats, so the respective risks are alleviated.

3.1 Safety Risks for Vehicular OTA Software Updates

As already mentioned in Section 2.2, remote installation of software updates, combined with OTA distribution means, provides several benefits to both OEM and vehicle owners. However, a number of safety issues might arise in contrast to updates installed at a service garage, where professional service personnel, for instance, can directly react on failed updates. Therefore, an OTA software update process has to foresee and to protect against several error situations and difficult conditions that could cause an update to fail. This especially important since failed updates might render an ECU a completely unusable, which might have safety implications for the whole vehicle, depending on the type of ECU. In the following subsection, we hence describe some exemplary safety risks for vehicular OTA software updates.

Improper software management – Especially partial or differential software updates are made-to-measure for a certain target software version. If these updates are applied to a different version, the control flow insertions, modifications or deletions may not make any sense or even damage working code. Also, there may exist interdependencies between a target software and other modules. Changing the one without updating the other would lead to non-functional code, for example due to linkage errors.

Insufficient network connectivity – It is typical that vehicles move in areas with low or no mobile network connectivity. Update package downloads need to be expected to become interrupted due to bad connectivity that may occur in tunnels, rural areas, or underground garages.

Unsafe situations – Even once the download has been completed, the update may not directly be installable, because the installation itself might take several minutes (see Section 2.3). The vehicle could be in an unsafe situation, where it relies on the availability of the target ECU.

Insufficient resources: At the target vehicle and/or ECU could be insufficient resources for the update installation, such as battery power, memory, or computation capacities.

Unexpected hardware issues – During the update, several errors could occur due to material fatigue, for instance caused by too many update write cycles into memory cells. Also, a crucial ECU could be subject to an emergency stop while an update is in progress, which would leave the update routine no possibility to complete the update.

User acceptance – An automated update management system may detect suiting situations for installing software updates, but the driver's time schedule and the intended vehicle usage by the driver cannot be anticipated. The driver understands the vehicle use in the near future the best and should not be constrained in the vehicle usage merely because of an update-in-progress.

3.2 Security Risks for Vehicular OTA Software Updates

For a systematic risk evaluation, it is necessary to identify the critical *assets* of a typical OTA update system setting like the one that is depicted in Figure 2. For each asset, the relevant aspects are then identified that need to be secured. These aspects are referred to as *security objectives*. Table 2 summarizes the assets, security objectives and identified threats.

Software update package – An update package contains all the necessary information to update the specific ECU software. For this reason, the most obvious attacks against an ECU base on tampering with these packages. In particular, the *authenticity* and *integrity* of update packages need to be protected to impede the import of forged and manipulated data, respectively. Also, replay attacks of old – back then valid and authentic – update packages need to avoided, thereby securing the *freshness* of update information. Furthermore, every update package carries intellectual property of the system manufacturer and therefore needs to be treated *confidentially* during the update process. The installation of an update packages also needs to be tracked and so, the *non-repudiation* of an update transaction needs to be secured.

ECU software – The ECU software represents the functionality of the ECU. This includes the core functionality as well as the applications that enable the system's use cases. It is crucial that the software's *integrity* remains unharmed, i.e., neither malicious, nor correct, but unauthorized code gets injected. Also, *denial of service (DoS)* attacks needs to be prevented to protect the *availability* of the system and the use cases.

Vehicle – Similar to the ECU software, the vehicle represents the functionality of the system, but on a more holistic level. Therefore, the security objectives *integrity* and *availability* also apply here.

Passenger – The passenger is the most important asset in every system that interfaces with human users. It is the transition from purely security-related to safety-critical considerations. Every physical harm or injury of the passenger has to be avoided, which means that the *integrity* of the passenger has to be secured.

Backend – The backend contains the whole functionality to generate software update packages. The injection of malicious code into the generation toolchain is a basic demand for software management systems. Therefore, it is out of scope of the special

OTA update scenario and is omitted here, but beyond that, the *availability* of the update service needs to be protected to keep the devices in the field up-to-date.

Table 2: Average update data transfer times on typical automotive bus systems.

Asset	Security Objective	Threat
Update Package	Authenticity	Forging of valid, but malicious update package
	Confidentiality	Disclosure of software IP
	Freshness	Replaying an old update package
	Integrity	Improper manipulation of update package
	Non-repudiation	Denial of update transaction by the backend or the ECU
ECU Software	Availability	Unavailability of the ECU functionality
	Integrity	Improper manipulation of the ECU functionality
Vehicle	Availability	Improper manipulation of the vehicle's functionality
	Integrity	Unavailability of the vehicle
Passenger	Integrity	Injury of passenger
Backend	Availability	Unavailability of Backend

Based on these threats, we can assemble a set of security requirements. By fulfilling these requirements, the regarded OTA software update system can be effectively protected against the threats; however, each threat can be realized in various ways, so that the concrete implementation of the security requirements depends on several different parameters. These include, for example, the expected attacker, existing security means, and available resources.

Requirement 1: Protection of transmission of update package – During the transmission from the backend to the vehicle, the update package needs to be secured with respect to its confidentiality, integrity, authenticity, and freshness.

Requirement 2: Protection of stored update package – After an update package has been received from the backend, it needs to be stored securely in the vehicle, before it can be installed. In most cases the target ECU cannot be used during the update process, which may significantly affect the vehicle's functionality. Therefore, an update package is stored until the vehicle is idle for a longer period of time, which is when the packages can be safely installed (see Section 2.3). Also, the internet gateway ECU of the vehicle, where the connection from the update distribution service is terminated, might not be the target ECU for which the update package was generated for. So, until the package can be installed, it still needs to be secured with respect to its integrity, authenticity, and freshness, as an attacker may gain access to the ECU, where the update package is stored and manipulate the package directly in the storage. In some cases, the confidentiality of the update package might also need to be protected in the connected vehicle;

this is not an essential part of an OTA update system and is therefore not regarded in this paper.

Requirement 3: Verification of update authorization – Even if the integrity and authenticity of the update package itself is protected, it needs in addition to be verified that the instance that had issued the update package has the necessary access rights and authorizations to install the update at the designated target device. For example, if the issuing instance at the backend is part of a complex infrastructure, it needs to be carefully verified, whether the entity may not only issue update packages at all, but also may do so for the addressed target ECU.

Requirement 4: Protection of the update transaction – In several cases, the provision and installation of an update package has to be traceable. For example, an update might be safety-critical and the OEM needs to track the installation coverage of a certain update. Also, some update packages might be erroneous and could cause an error and the vehicle owner needs to prove that the error results from an official, i.e., authentic, update package. Therefore, the installation of an authentic update package has to be documented by both, the OTA backend and the target ECU.

Requirement 5: Protection against overloading the backend – Update packages that fix safety- or security-critical issues in ECUs need to be readily available to the target devices. An OEM needs to make sure that the distribution backend is hardened against DoS attacks, so that this availability is not impaired.

4 Safe & Secure OTA (SOTA) Software Updates

In order to thwart the safety and security risks identified in Section 3, we extend the original OTA architecture with effective protection measures to make it a safe and security over-the-air (SOTA) software update architecture.

Hence also the original OTA update procedure described in Section 2.3 will be extended as shown in Figure 6 and shortly described in the following.

11

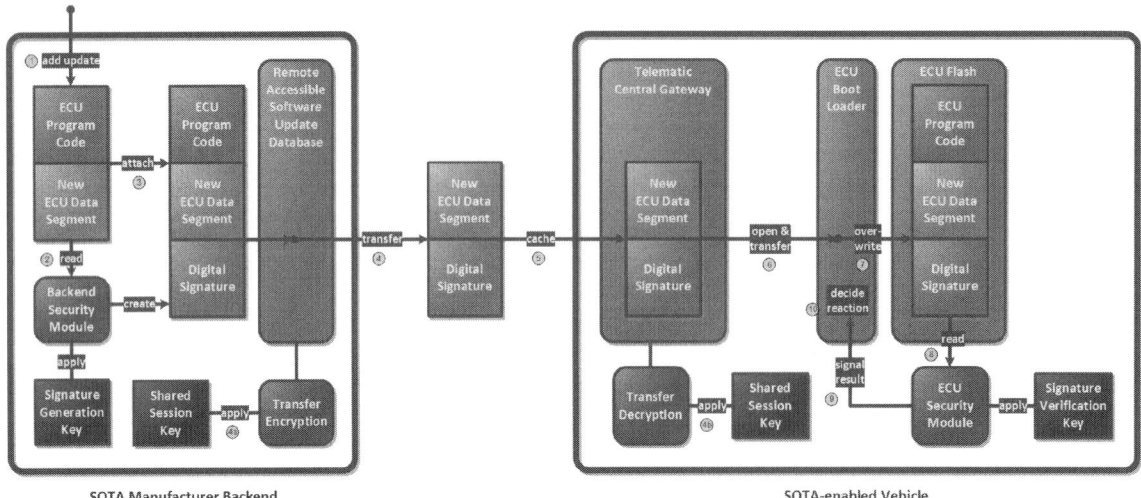

Figure 5: Functional overview of a SOTA software update.

After the OEM has approved a new own or a new third party ECU software (1), the complete ECU software package consisting of program code and program data will be signed by the backend security module using the OEM signature generation key with a proper cryptographic signing algorithm such as RSA-2048 or ECC-256 (2). The resulting digital signature is then attached to the ECU software package (3) and stored the OEM software data base to make it available for remote download by corresponding vehicles. When a compatible vehicle now remotely connects with the OEM backend, it can wirelessly download (4) and cache (5) the new software package together with its digital signature at the CGW of the SOTA-enabled vehicle. The transfer channel between backend and vehicle can optionally also become protected against unauthorized interceptions by encrypting also the communication channel using a pre-shared session key together with a proper cryptographic encryption algorithm such as AES-128 (4a, 4b). If the data is completely received at the vehicle, and the CGW has checked all necessary access rights, user approvals, and vehicle safety conditions, it opens the corresponding target ECU and transfers the complete software package together with the digital signature to the ECU bootloader (6), which in turn will write it directly into the ECU flash memory usually while overwriting the old software package at the same time (7). Afterwards, the ECU security module, realized for instance by a proper automotive hardware security module such as SHE or HSM, reads the new flash memory chunk by chunk and cryptographically verifies it for its authenticity and against any unauthorized manipulations using the corresponding OEM signature verification key (8). Based on the verification result that the ECU security module signals to the bootloader (9), the bootloader then decides about the next steps to be taken such as reboot and execution in case of success or emergency mode and error log entry in case of an update failure.

A digital signature and an optional transfer encryption alone, however, are not sufficient to make the OTA update a fully safe and secure SOTA update. As shown in Figure 6, there are a few more safety (violet) and security (green) protection measures necessary and available to increase the dependability of the SOTA update accordingly.

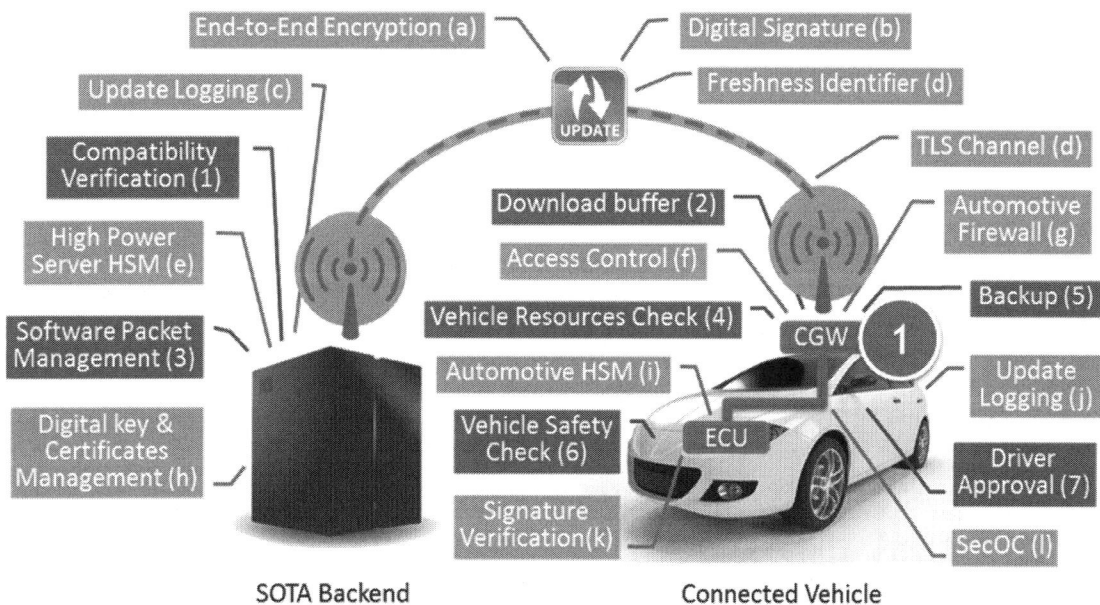

Figure 6: Safety (violet) and security (green) measures for SOTA update architectures.

Hence in the following a quick overview about additional safety and security measures is given.

4.1 Additional SOTA Update Safety Protection Measures

The compatibility verification (1) ensures that only software verified to be fully compatible and compliant with the requesting vehicle and its actual software configuration will be allowed for SOTA update installation (vs. Section 3.1: Improper software management). Otherwise it will deny the installation and propose necessary steps to pass the compatibility verification (if any and if possible).

The download buffer (2) ensures that the software update can be received completely even if the mobile connection becomes interrupted during the mobile data transmission (vs. Section 3.1: Insufficient network connectivity).

Based on the VIN given together with the current vehicle software configuration, the software package management (3) checks for new vehicle software that is potentially available for a SOTA update installation (vs. Section 3.1: Improper software management). It further checks and ensures that all necessary prerequisites and interdependencies that might be necessary are solved (first) to install a new software update (e.g., certain additional software library).

Based on the given software update size and target, the vehicle resources check (4) ensures that all necessary resources are (and remain) sufficiently available before the update will be finally installed. Therefor it checks for instance for sufficient battery power and temperature over the course of time, sufficient on-board data transmission capacity, or sufficient target memory (vs. vs. Section 3.1: Insufficient resources).

Since the software update process usually happens outside professional garages, it is further possible to have a backup memory (5) installed at the vehicle that saves the original software before it becomes updated to be always able to fall back to the original software again in case the software update causes any failures (vs. Section 3.1: Unexpected hardware issues).

Similar to the vehicle resources check, the vehicle safety check (6) ensures that the vehicle and all vehicle parameters are (and remain) in safe conditions before the update will be finally installed. Therefor it checks for instance the vehicle being in a parking situation (vs. Section 3.1: Unsafe situations).

Once the vehicle has been checked for sufficient resources and proper safety conditions, the driver approval check (7) finally tries to obtain the obligatory acceptance of the authorized vehicle owner/driver (who will be first authenticated to verify his authorization accordingly) to finally apply the SOTA update. The driver therefore gets also a forecast about the total the time needed for installing the update, in order to be able to assess if he can let have the vehicle unutilized for this time period (vs. vs. Section 3.1: User acceptance).

4.2 Additional SOTA Update Security Protection Measures

As already described at the beginning of Section 4, the transfer channel between backend and vehicle can optionally become protected against unauthorized interceptions by encrypting it using a pre-shared session key together with a proper cryptographic encryption algorithm at the backend and at the CGW. Moreover, even the complete end-to-end transfer channel between backend and actual target device can become protected (a) against unauthorized interceptions by encrypting the update software itself having for instance a pre-shared session key and a proper cryptographic encryption algorithm securely available at the backend and the target device itself.

The application (b) and verification (k) of the digital signature is described at the beginning of Section 4.

Update logging at the SOTA backend (c) and at the vehicle (j) ensures mutual proof that a certain update has or has not been applied at a certain time by a certain role etc. realized usually with a cryptographic approach based on the concept of non-repudiation.

Freshness identifier (d) ensures the current update is really the last up-to-date update and hence protects especially against replay attacks. Typically this is realized by a cryptographically protected monotonic counter.

High power server HSM (e) represents a high-performance, industrial-scale hardware security module (e.g., SecurityServer CSe from Utimaco) that ensures that all critical cryptographic data (e.g., root signing keys) and operations (e.g., digital signing) are hold and executed always in a physically shielded environment only but with full hardware acceleration.

Access control at vehicle (f) enforces that only authenticated and authorized roles (e.g., driver) can execute authorized operations (e.g., overwrite) on certain data objects (e.g., ECU software) based on mandatory access control mechanisms usually involving two-factor authentications and verification of physical presence.

Automotive firewall (g) ensures that only successful authenticated and fully authorized parties (e.g., servers, devices) can communicate with the vehicle using classical black and white list rules but also more advanced firewalling mechanisms such as packet filtering, heuristics, or intrusion detection together with active response measures.

Digital key and certificates management (h) represents a large-scale credential management system capable amongst other for secure creation, storage, update, distribution, logging, deactivation from millions of all kind of cryptographic assets such as asymmetric and symmetric keys of different sizes and representations (e.g., ECC-256 keys), all kinds of digital certificates (e.g., embedded X.509), all kinds of access authorizations (e.g., password hashes), reference values (e.g., platform configuration values) etc.

Automotive HSM (i) represents an automotive-capable hardware security module (e.g., SHE or Bosch HSM) that ensures that all critical cryptographic data (e.g., pre-shared keys) and operations (e.g., encryptions) are hold and executed always in a physically shielded environment not always but often also including some hardware acceleration.

5 Conclusion & Outlook

In this paper, we focused on the safety and security aspects of over-the-air (OTA) software update architectures for advanced driving assistance systems. We described the current state of the art, discussed the safety and security implications of OTA software update systems and presented a clear description of a safe and secure over-the-air (SO-

TA) software update process and a general SOTA software update system architecture that fulfills the identified safety and security requirements.

In the near future, OTA software updates will become the standard way for electronic maintenance and support for virtually all new-developed vehicles. It will be used to quickly and efficiently ensure safety and security of modern connected vehicles over the complete lifecycle. Thereby, it will increase the coverage of up-to-date ECU software and the convenience of vehicle owners at the same time, as many costly visits at service garages will become unnecessary. Potential safety and security vulnerabilities in complex ECU software systems can become get thwarted efficiently and effectively. Indeed, the introduction of SOTA software updates itself might also introduces some new risks regarding safety and security that cannot become completely neglected. However, the huge benefits for vehicular safety and security that are newly gained by the introduction of SOTA updates outbalance these additional minor risks by far.

Last but not least, as soon as a SOTA architecture has been reliably established, it will be a trailblazer for many new use and business cases for connected vehicles. Standing on the shoulder of giants, remote diagnosis and feature activation, mobile payment, automotive data synchronization services and many more will rely on the very techniques and mechanisms that enable a secure over-the-air software management.

Bibliography

[1]   BMW Presse- und Öffentlichkeitsarbeit: BMW ConnectedDrive. Navigationskarten-Updates per Mobilfunk und integrierte Spritpreissuche. https://www.press.bmwgroup.com/deutschland/article/attachment/T0177750DE/2 62435, Presse-Information, April 10, 2014, retrieved on February 24, 2016.

[2]   Damon Lavrinc, Wired: In Automotive First, Tesla Pushes Over-the-Air Software Patch, http://www.wired.com/2012/09/tesla-over-the-air/, September 24, 2012, retrieved on February 24, 2016.

[3]   Reuters: BMW fixes security flaw in its in-car software, http://www.reuters.com/article/bmw-cybersecurity-idUSL6N0V92VD20150130, January 30, 2015, retrieved on February 24, 2016.

[4]   Redbend Ltd, White Paper: Cost Effective Updating of Software in Cars, February 2015

Multi-functional open-source simulation platform for development and functional validation of ADAS and automated driving

Lei Wang[1], Timo Vogt[1], Jan Dobberstein[2], Jörg Bakker[2], Olaf Jung[1], Thomas Helmer[1], Ronald Kates[3]

1   BMW Group, 80788 München; +49-89-382-58984, lei.wl.wang@bmw.de
2   Daimler AG, 71059 Sindelfingen
3   REK Consulting, 83624 Otterfing

© Springer Fachmedien Wiesbaden GmbH, ein Teil von Springer Nature 2018
R. Isermann (Hrsg.), *Fahrerassistenzsysteme 2016*, Proceedings,
https://doi.org/10.1007/978-3-658-21444-9_10

Introduction

In modern vehicles, Advanced Driver Assistance Systems (ADAS) and automated driving functions are increasingly playing the role of a co-pilot, supporting the driver in complex or dangerous situations by applying preventive strategies.4 These strategies include warnings, enhanced braking assistance, and automatic interventions to increase road safety. As discussed in [2], these strategies can help to avoid collisions, or – in case of inevitable accidents – mitigate injury severity. Individual customers, automotive manufacturers and suppliers, public officials, regulatory agencies, insurers and consumer protection organizations are important stakeholders in development, deployment, and assessment of safety relevant vehicle functions.

A key issue for all stakeholders is assessment of ADAS safety performance, including quantification, not only of protective potential, but also of potential adverse consequences, such as false-positive responses. For example, in the case of autonomous emergency braking systems, adverse consequences could be associated with induced rear-end collisions.

Safety criteria play a primary role in development, approval, and deployment of novel vehicle systems, ranging from advanced driver assistance to automated driving. With the advent of ADAS, the complexity of safety assessment has vastly increased as has been addressed in public research projects [5, 10]. Ideally, it would be possible to make quantified predictions of traffic safety of new systems so that all stakeholders – despite their partly divergent interests – can agree on validity and reproducibility of these predictions [6]. This goal requires agreement on precisely defined traffic safety metrics appropriate to the particular context, as well as consensus on valid methodologies for their assessment [3]. Furthermore, a common platform that can independently be used by all stakeholders to ensure transparency, comparability, and plausibility of the assessment results, is indispensable. This paper presents a novel open-source software platform, in order to meet these methodological and practical requirements for assessment of assisted and automated driving functions.

---

4 Since ADAS and automated driving functions rely on comparable technology and operational principles (i.e., sensing, warning and intervention), automation could be considered as an enhanced and continuously operating kind of ADAS. In this context, the term ADAS refers to both assistance and automation functions.

Methodology of safety performance assessment

Conventional testing methods, as currently used, include hardware-in-the-loop procedures (e.g., for sensor / algorithm testing), testing of technical and human factors (e.g., driving simulator, test track, test rigs), and methods based on real-traffic testing (e.g., controlled studies, field operational tests, observational studies). Each of those methods provides an important insight into the safety effectiveness of a system on a particular small subset of processes influencing the likelihood and severity of an accident. For complex active safety problems, conventional testing methods are rarely sufficient to cover all relevant processes and make sound estimates of the overall safety effects in traffic.

An important challenge is to assess field effectiveness with adequate validity: Any key aspect that is assessed with low validity will disproportionately compromise the validity of the entire assessment process, given sufficient sensitivity of that particular aspect. Simulation approaches based solely on reconstruction of real-world crashes (e.g., using the GIDAS-based Pre-Crash Matrix (PCM) [9] suffer from several shortcomings: First, the spectrum of sampled scenarios is limited to those in the database. Secondly, they do not provide an estimate of adverse consequences, such as false-positive system reactions; thus key performance statistics, such as sensitivity and specificity cannot be estimated.

In this context, a novel and enhanced assessment approach is being discussed and refined within the international P.E.A.R.S. initiative (Prospective Effectiveness Assessment for Road Safety). A basic scheme of this approach based on multi-agent based, stochastic traffic simulation is illustrated in Fig. 1.

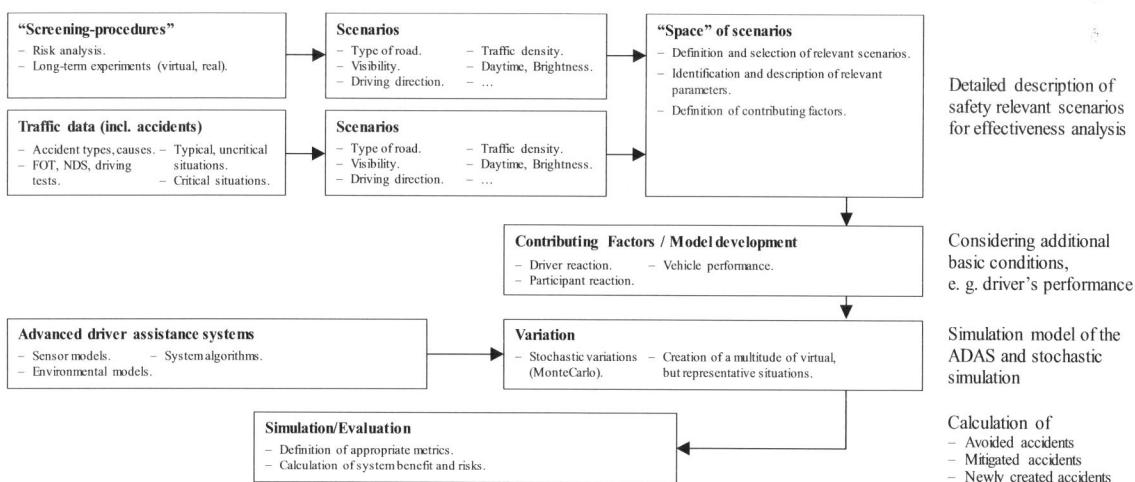

Fig. 1: Schematic description of multi-agent stochastic effectiveness assessment approach (following [4]).The general approach is to incorporate models of traffic scenarios including road layouts, cognitive and behavioral models of traffic participants, environmental conditions, and models for vehicles and their embedded safety functions into a simulation platform [3, 8]

In a first step, target scenarios that could be positively affected by a proposed system need to be identified. This stage usually involves screening procedures based, e.g., on risk analysis and subsequently detailed description of the target scenarios on the basis of traffic and accident data.

Once relevant target scenarios for the assessment have been identified, in a second step, statistical exposure models are determined in order to generate representative samples for the simulation. Statistical exposure models provide distribution functions of traffic context variables that could influence the system effectiveness in a certain scenario. Relevant traffic context variables can include environmental factors (e.g., lighting conditions), infrastructural characteristics (e.g., road curvature), macroscopic traffic state (e.g., traffic density), and microscopic features (e.g., gap distribution between traffic participants). An appropriate exposure model should provide a representative sample of instances in which the system in question could be activated. Thus, representative samples can be generated by repeatedly drawing instances of the scenario from this exposure model.

The third step focusses on reconstruction or modelling all key processes within those scenarios that possibly contribute to accident risk and that can be influenced by an ADAS within a target scenario. These processes include psychological, cognitive, physiological, and physical phenomena. Statistical uncertainties due to inter- and intra-individual variations in driver behavior (e.g., reaction speed) and system behavior (e.g., sensor performance) need to be considered as stochastic parameters of the models.

The last preparatory step of the assessment is definition of safety performance metrics. An appropriate metric for safety performance of any ADAS must take into account reduction of accidents as well as their injury outcomes in the addressed scenarios. However, since active safety design usually involves dilemmas and trade-offs between benefits and risks due to the inherent uncertainty of predicting the (near) future of events, balanced results must consider both benefits and risks. In other words, the metric should include "true-positive", "true-negative", "false-positive", and "false-negative" system response rates.

Implementation of this novel "holistic" methodology for safety effectiveness assessment poses high demands on an appropriate software solution, which has to handle large numbers of relevant scenarios, high complexity of the models, stochastic processes, and a multi-agent-structure. Furthermore, to ensure transparency, comparability, and standardization of future assessments, it should ideally be easily available for all stakeholders of traffic safety. Recently conducted market research has revealed that none of the currently available simulation tools can meet all these requirements. As consequence, a novel open-source platform, openPASS (**Open P**latform for **A**ssessment of **S**afety **S**ystems), is in development. This platform, based on an initiative of BMW and Daimler and in accordance with the P.E.A.R.S. approach, is designed to apply this holistic methodology of ADAS safety assessment in a comprehensive way.

The openPASS Software Platform

Objectives of openPASS

Many individual objectives relevant for openPASS are well known within the context of traffic and accident simulation. However, the combination of requirements poses a considerable challenge. One key aspect is performance: The performance must be sufficient to provide the required statistical precision within a reasonable time-frame, i.e., calculating large numbers of scenarios. Another is flexibility: scenarios should be easy to adapt, and parameters within these scenarios should allow automatic variations in order to represent "realistic" stochastic conditions of traffic participants, technical systems, or environmental conditions.

The openPASS platform should be open and transparent. It must enable different types of usage (such as desktop PC simulation, cloud-computing, interface to driving simulator), but should also provide a secured access in order to use protected data or knowledge. Beyond the use case, the software design and architecture should encourage usage in, e.g., third-party funded research projects: to this end, the adaptation to new interfaces or implementation of scientific insights in new or existing modules should be possible. Ideally, openPASS should allow a wide range of modifications of the original source code.

Furthermore, the following aspects are of key importance and need to be considered within the platform development:

Performance: The performance requirements of openPASS aim for a very high number of runs to be conducted much faster than real time. The simulation core of openPASS is designed to meet this requirement and openPASS provides several options for enhanced performance, such as distributed computing.

Input data: In general, traffic accidents or conflicts are rare events in terms of occurrence. As a consequence, they are hard to capture in statistically significant quantities, e.g., by Naturalistic Driving Studies or Field Operation Tests. However, there are principally two approaches to assess effectiveness, although data is sparse: first, by re-simulating reconstructed pre-crash trajectories leading to accidents (as discussed above), or second, by generating virtual traffic scenarios which eventually lead to conflicts or accidents. These types of "input data" or starting conditions differ largely in terms of format, accuracy and level of detail and general methodological consideration (see above).

Stochastic variation: Software tools currently used for design and assessment of ADAS often have their focus on vehicle dynamics and / or certain defined scenarios. Subsamples of the wide range of all potentially occurring conditions are later addressed by testing the function in hardware under real-world conditions. Whereas system effects on all potential traffic situations as well as accidents cannot be observed by field testing alone (see above), virtually generated scenarios are able to cover the "space of scenarios and conditions" reliably, efficiently and in fast testing cycles. Key stochastic parameters are drawn from appropriate model statistical distributions. The shapes, parameters (locations, spread), and multi-variate characteristics of these distributions are derived from different fields of research, e.g., accident statistics, driver behavior studies or field testing data.

Modular architecture: The general architecture of openPASS should enable both convenient use and high performance. To this end, the whole project is designed as a "white box", i.e., the source code is freely available. However, the modularity is not limited to models, e.g., sensors or driver assistance system models, but also covers other software components, such as GUIs or module interfaces.

Flexibility & transparency: mainly, flexibility is provided by modularity within the architecture. In addition to that, interfaces and basic functionalities of the simulation core have to be transparent to ensure a maximum of understandability and further development.

Easy-to-use GUI: A software platform with a broad user base needs an easy-to-use GUI, which allows non-programmers (e.g., system developers or regulatory bodies) to use the tool. Therefore, at least all settings necessary for a standard assessment, e.g., selection of scenarios or modification of ADAS parameters, should be easily accessible.

Sustainability: The goal of sustainability means choosing a development environment and deployment model for the software that includes unlimited and free access for a wide range of users. A state-of-the art license model provides an option for future initiatives to use the software and adapt it to their needs.

Architecture, modules, and interfaces

The architecture of openPASS links different modules in order to create traffic situations, which may or may not result in conflicts or accidents. The architecture of the simulation core is structured in one "master" and multiple "slave" modules; enabling parallel processing. The master initializes the scenario, agents, and conditions for the simulation, defines single "simulation jobs", and coordinates the slaves.

The slave conducts a specific simulation run based on the scenario and the specific simulation task. This could mean that it creates a "new" generic scenario from an exposure model, in case trajectories are not given from a database. When the defined ending conditions are reached in a run (e.g., collision or a certain covered distance or simulation time), the slave hands over the results of its run to the master and terminates itself. The master adds all information from individual slaves to a general database and conducts overall statistical post-processing.

Other relevant information about scenario, agents, and conditions are provided to the simulation core by defining further modules in order to implement the preparatory steps of the assessment as described before. For example, the "starting conditions module" implements the scenario model of interest which includes sub-models of, e.g., the road environment, staring conditions for traffic or characteristics of other traffic participants. The sub-models are modularly designed as well, e.g., a driver assistance system model consists of software modules representing components such as sensors, actuators and interfaces to the vehicle dynamics model. Each of the modules in the software architecture can be exchanged or enhanced as well. Finally, it should be feasible with openPASS to model all aspects that could be relevant to the traffic conflict or accident outcome – depending on the actual evaluation objective.

openPASS should be used by different stakeholder in vehicle safety, and therefore, maximally interoperable within current software landscape. Hence, it will contain various interfaces and the possibility of adding required interfaces. One example is OpenDRIVE as a road description – the current de-facto standard format to exchange scenarios between driving simulators. Consequently, openPASS is designed to allow (basic) data to be exchanged between driving simulators experiments and traffic conflict simulation (or others) without conversion from one "world" to the other.

Harmonization

While some input data or settings are specific to the evaluation context and objective (e.g., assessment of the safety performance of a specific sensor configuration), certain models are based on common laws of physics or on empirical facts. For this reason, these models need to be standardized, which means that the development, validation, approval, and application need to be harmonized. By providing basic harmonized models, openPASS facilitates integration of various approaches. For example, a two-track vehicle model is a generally needed module for many use cases. While the specific vehicle dynamics model can be further specified, the key principles – how detailed mechanisms, such as drift or steering torque, are incorporated – can be further specified to account for certain vehicle characteristics. The initiative behind openPASS aims at covering a wide range of state-of-the-art knowledge from related research fields (such as traffic psychology, traffic flow management, and traffic sciences) by integrating these findings into basic modules of the software.

Eclipse as open-source ecosystem

Open-source software like Python or R has spread in recent years because of large and active communities of software developers, fast feedback regarding newly implemented features, high connectivity in internet forums, and a "low entry hurdle" – no immediate cost. Moreover, the aforementioned objectives of modularity, flexibility, and transparency are basically the key principles of open-source initiatives. They can hardly be achieved by closed-source software, since even very advanced settings or interfaces require knowing beforehand which parts of the code could be potentially changed. Given the knowledge of how to adapt the code, independent changes or additions can be made to open-source code.

The interdisciplinary community in vehicle-safety has a strong history in tool development and "in-house" solutions. Easy and free-of-charge access should allow all interested institutions to quickly adapt to the software and to communicate in a common language, whereas now, simulation results are closely linked to the software environment used for an assessment. Open-source assessment software will enable new synergies and the exchange of knowledge between different stakeholders and research areas. Hence, it is well suited for this community. Given the task openPASS is facing – a realistic traffic model right down to the specific conditions of very rare accident events – this intensified exchange is much needed.

From a corporate IT department point of view, the actual use and further deployment of open-source software might not appear as easy as buying and using proprietary software. It may often require a stronger involvement of the IT department. Even given the flexibility goals of this platform project, the additional short-term efforts in terms of further code development may appear greater than the long-term benefits, e.g., for building, testing and adapting the raw software to the specific application. Especially for companies, there are further hurdles in terms of liability, stability and cross-company support for such open-source solutions. Hence, the professional use of open-source software is more convenient if the software is well established, already delivering performance and easy-to-use of comparable solutions – or superior quality – to those of proprietary software. This is clearly not the case for the novel approach of openPASS, which requires new software and foresees the need to adapt it following progress in traffic safety related sciences.

In order to reconcile the advantages and pragmatism of open-source communities and the software development needs of companies, the Eclipse Foundation offers the concept of so-called Industry Working Groups [1]. Eclipse is a community for individuals and organizations who wish to collaborate on commercially-friendly open-source software. The Eclipse Project was originally created by IBM in November 2001 and supported by a consortium of software vendors. The Eclipse Foundation was created in January 2004 as an independent not-for-profit corporation to act as the steward of the Eclipse community. In addition to the support of "classic" open-source, Eclipse Working Groups allow for organizations to collaborate in the development of new innovations and solutions, e.g., for the "Internet of Things" or for storing and handling large amount of measurement data like openMDM [7].

8

The concept of the Eclipse working group process is as follows:

- The Eclipse Foundation provides a generic infrastructure that can be adapted to the specific purpose. This includes standard processes and bylaws[5], the Eclipse development environment, and membership levels accounting for different roles. Most of all, the working group concept aims at linking Eclipse projects that develop code with interested parties financing these activities.
- Each working group describes its general strategy, specific intentions, and goals; it also defines different roles, e.g., driver member investing in code development, or service provider members, i.e., companies with experience in developing software for related use cases.
- Once a working group is in place, related Eclipse projects are started to implement its goals. A project leader assures that the Eclipse development process is executed and the code is managed within the Eclipse repositories. This is accompanied by working group committees focusing on requirements management, architecture compliance, and quality assurance. These bodies are staffed with personnel of the working group members involved in the specific topic.
- Open-source software is developed by professional software engineers, providing high quality, while the involvement of future users ensures high relevance for the objectives of the working group.
- All meeting minutes and documents related to the working group are public as well.

Due to the Eclipse Public License and the open ecosystem of the Eclipse foundation, various projects might be forked from the original code, and further members can join the Working Group if they accept the charter and its rules. They may join existing projects or start new ones, as desired.

For openPASS, this approach provides flexibility, creativity, transparence, and openness to a very large extent – regarding members and code as well as access to the scientific community. On the other hand, the processes of the Eclipse foundation provide a stable and organized environment, ensuring a high quality standard and professional project management. The Eclipse working group "openPASS" was officially founded in March 2016. The first deliverable of the related Eclipse project "SIM@openPASS", publishing the core parts of the platform, is expected by the end of 2016.

---

5   https://eclipse.org/org/documents/

The broader perspective regarding rating processes

For advanced testing and rating procedures designed to deliver results accepted by all stakeholders, there is an urgent need for integration of different assessment instances and their results into stochastic simulation for a "holistic" rating. To this end, a concept of extensive knowledge synthesis is required [3]. This concept should be based on cooperative contributions from neutral (scientific) institutions, automobile industry, and established test institutes, representing the interest of customers. Within this concept (Fig. 2), advanced models of the key processes of certain target traffic scenarios and the corresponding models are developed and validated by neutral (scientific) institutions in order to maximize the validity and transparency of the simulation-based assessment procedure.

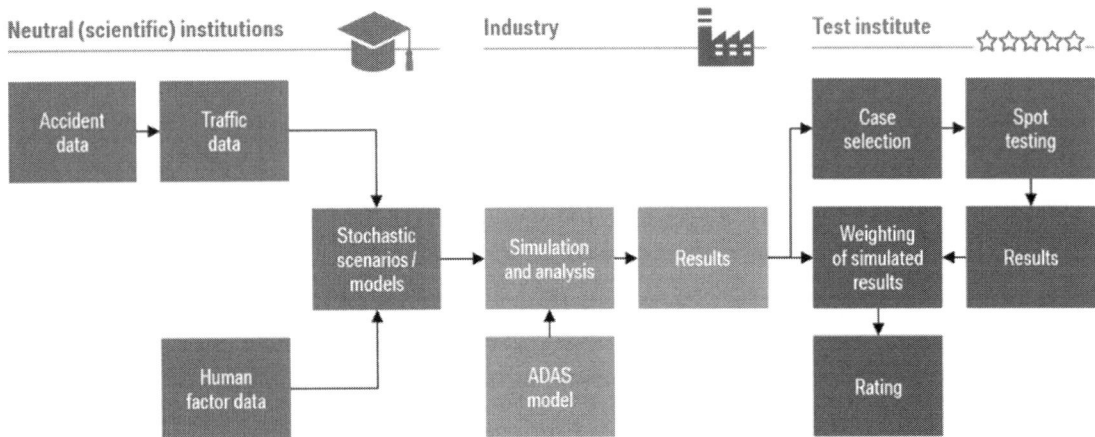

Fig. 2: Knowledge synthesis and cooperative safety effectiveness assessment of ADAS.

The manufacturer simulates system effects using agreed-upon and harmonized models and then provides the results to a rating organization. The rating organization could, for example, perform real-vehicle tests in randomly selected individual situations from simulated samples and apply a weighting factor based on the real-vehicle test results to the simulation results, which in the end provide the basis for the actual rating.

Standardization of methods and models

While openPASS is a software platform that allows stakeholders to easily assess ADAS, it also requires a defined, established and accepted methodological process in place. As stated above, there is an urgent need for advanced testing and rating procedures capable of prospectively quantifying the safety performance of safety system – according to a metric accepted by all stakeholders – regardless of whether the engineering approach involves passive elements, active systems, automated functions, or integrated combinations of all these.

Stochastic simulation has an immense potential to enable more meaningful and realistic rating procedures, supported by knowledge synthesis, as a key component. While this proposed scenario involves all relevant stakeholders (e.g., neutral scientific institutions, industry partners as well as test institutes) it obviously has to rely on a standardized procedure. Besides defined technical interfaces like the ones defined within the openPASS Project, an established and accepted method and process between all involved process partners are needed. The P.E.A.R.S. Group (Prospective Effectiveness Assessment of Road Safety), which includes a substantial number of international key players, has established an open platform to discuss standards and procedures regarding the methodology to assess ADAS. Since its founding in 2012, more than thirty European and non-European partners have shown their commitment to contribute to definition of a standard for prospective safety assessment incorporating simulation.

P.E.A.R.S. Group

This informal working group consists of members of various institutions. The working group collects and exchanges best practices and experiences. The group is open and explicitly welcomes new members. At present, no official project form or financing model is established; this is subject of further work.

A key objective of this open working platform is the preparation of a worldwide standard for the evaluation of systems, which has been discussed and finally accepted by all relevant stakeholders. The focus is on a common and accepted understanding of the methodology used, independent of specific technical realizations. The current group agrees that the harmonization and standardization will be a continuing and iterative process, which will cover more and more use cases over the years to come. The advantage of the current actions lies in the concentration of forces regarding available knowledge as well as research into new issues, particularly those with common interests independent of market place competition.

The group presented itself for the first time in public during the ESV Conference in Gothenburg in 2015. The spirit of the group was and still is well accepted by all participants, and the outcome is regarded as beneficial, reflecting a common understanding when concerning the assessment of (new) ADAS. However, P.E.A.R.S. does not have binding character amongst stakeholders, governments or consumer protection agencies. In order to achieve a more binding character, the next logical step is to establish the methodology within a formal standardization body, such as ISO.

ISO Working Group 22 SC36 WG7

Eventually OEMs, suppliers, and governments as well as consumer advocates require a reference standard. Although an ISO standard has no direct legal status, it is an international and well accepted process that can describe the assessment of ADAS as a "state-of-the-art" process and can be referred to in a regulatory environment.

Therefore an ISO standard – as stated above – is the next logical step to establish an even more official and well-accepted method for assessment and analysis of ADAS. Independently of the P.E.A.R.S. Group, the ISO Working Group 22 SC36 WG7 has defined a new work item "Prospective Effectiveness Analysis", which is now linked to the P.E.A.R.S. initiative.

The goal within the ISO Working Group, supported by members of the P.E.A.R.S. initiative, should be to set standards regarding the methodology as well as regarding the minimum requirements for ADAS. This process is underway and will be an ongoing task for the future.

openPASS – summary

Key issues regarding assessment of ADAS safety performance and the needs for harmonization have been outlined in this paper. In line with the ongoing discussion of the international P.E.A.R.S. initiative (Prospective Effectiveness Assessment for Road Safety), further integration is needed in terms of a standard software tool, allowing multi-agent based, stochastic traffic simulation.

Given the objectives – performance, modularity, and transparence – a novel open-source platform, openPASS, is in development, based on an initiative of BMW and Daimler. This open-source initiative makes use of the infrastructure of an Eclipse working group. This group links industrial working group members who are defining requirements and financing development packages with Eclipse projects dedicated to open-source code development.

Invitation to join

Finally, openPASS – the core team of the Eclipse working group – encourages involvement of all interested stakeholders, e.g., as future users or driver members. What could be your role?

For further details on approach, goals, and code, please visit the related websites of Eclipse, the working group or the project.

To use openPASS: just download the code from the Eclipse filer system. The first official version of the platform "SIM@openPASS" will be officially released by the end of 2016.

To adapt and modify openPASS: in principal, just download the code and adapt it as you like, and let the project leader know your improvements, so it may become part of further versions of the software. You could also start a new project, forking the code.

Contribute your knowledge to the ecosystem: while the previous options aim at further development or modification, the actual content in terms of models need to be developed as well. This could be an additional Eclipse project or could be part of a project independent of Eclipse using openPASS and, e.g., creating additional modules.

Become a member of the working group: for entities such as companies or institutions, it could be of interest to become a working group member – please contact the authors in order to identify the role suited for you.

References

[1] Eclipse. https://www.eclipse.org/org/workinggroups/industry_wg_process.php, as of February 11, 2016.

[2] T. Helmer. Development of a Methodology for the Evaluation of Active Safety using the Example of Preventive Pedestrian Protection. Number ISBN 978-3-319-12888-7 in Springer Theses. Springer, 2015.

[3] T. Helmer, L. Wang, K. Kompass, and R. Kates. Safety performance assessment of assisted and automated driving by virtual experiments. In *2015 IEEE 18th International Conference on Intelligent Transportation Systems*, pages 2019–2023, 2015.

[4] K. Kompass, T. Helmer, L. Wang, and R. Kates. Sicherheitsbewertung von automatisierten Fahrfunktionen. *AUTOMOBIL-ELEKTRONIK*, (ISSN 0939-5326):20–22, 11-12 2015.

[5] V. Labenski, J. Dobberstein, and T. Schlender. Ur: Ban ka-wer: Accident data analysis and pre-crash simulation for the configuration and assessment of driver assistance systems in urban scenarios. *Berichte der Bundesanstalt fuer Strassenwesen. Unterreihe Fahrzeugtechnik*, 2015.

[6] P. Luttenberger, J. Bakker, E. Tomasch, R. Willinger, C. Mayer, N. Bourdet, C. Ewald, and W. Sinz. Method for future pedestrian accident scenario prediction. *Transport Research Arena 2014, Paris*, 2014.

[7] openMDM. www.openmdm.org, as of February 11, 2016.

[8]   Y. Page, F. Fahrenkrog, A. Fiorentino, J. Gwehenberger, T. Helmer, M. Lindman, O. op den Camp, L. van Rooji, S. Puch, M. Fränzle, U. Sander, and P. Wimmer. A comprehensive and harmonized method for assessing the effectiveness of advanced driver assistance systems by virtual simulation: The p.e.a.r.s. initiative. In *24th International Technical Conference on the Enhanced Safety of Vehicles (ESV 2015)*, number 15-0370, 2015.

[9]   A. Schubert, C. Erbsmehl, and L. Hannawald. Standardized pre-crash-scenarios in digital format on the basis of the vufo simulation. In *5th International Conference on ESAR "Expert Symposium on Accident Research"*, number F 87 in Fahrzeugtechnik. Bundesanstalt für Straßenwesen, Fachverlag nw, 2013.

[10] M. van Noort, T. Bakri, F. Fahrenkrog, and J. Dobberstein. Simpato-the safety impact assessment tool of interactive. *Intelligent Transportation Systems Magazine, IEEE 7 (1), 80-90*, 2015.

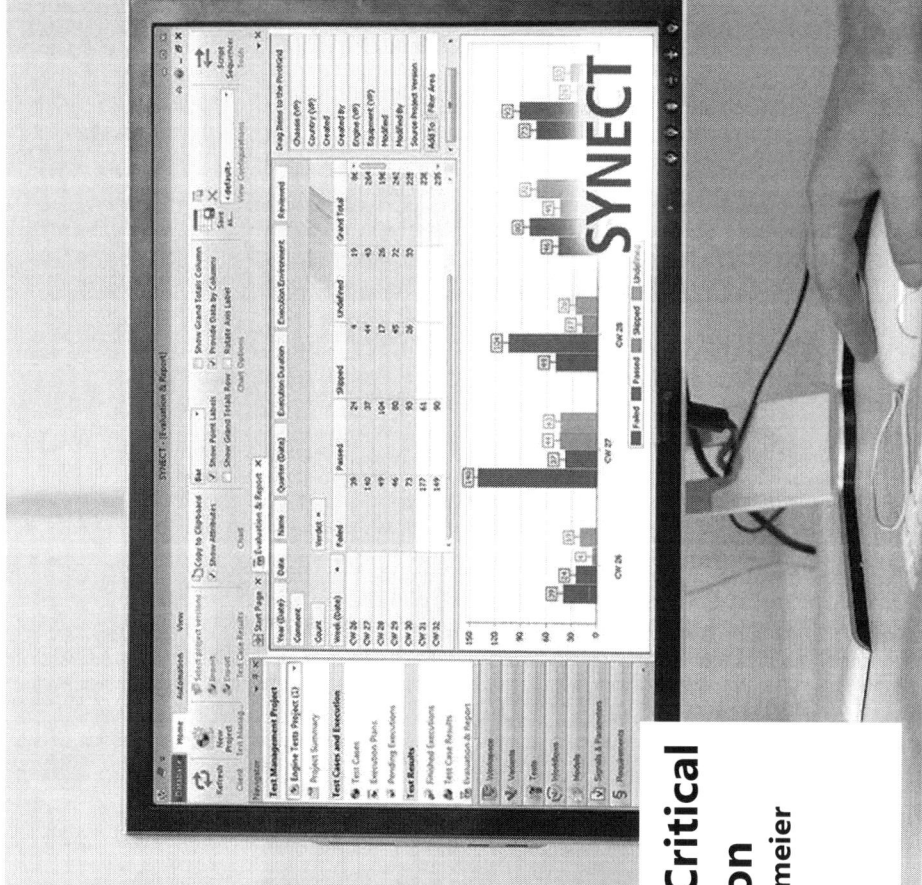

Embedded Success **dSPACE**

Integrated Tool Chain for Testing Safety-Critical Assistance Systems by Means of Simulation

Michael Beine, Claudia Hollmann, Gregor Hordys, Andre Rolfsmeier

14.04.2016

dSPACE GmbH · Rathenaustr. 26 · 33102 Paderborn · Germany

1

Vision of Accident-free Driving

European Commission

- Autonomous emergency braking (AEB) and lane departure warning (LDW) mandated for commercial vehicles

Euro NCAP

- Active safety will be part of the overall rating for passenger cars

  - 2014: AEB and LWD
  - 2016: AEB with pedestrian recognition
  - 2018: AEB with cyclist recognition
  - 2020: Junction Assist

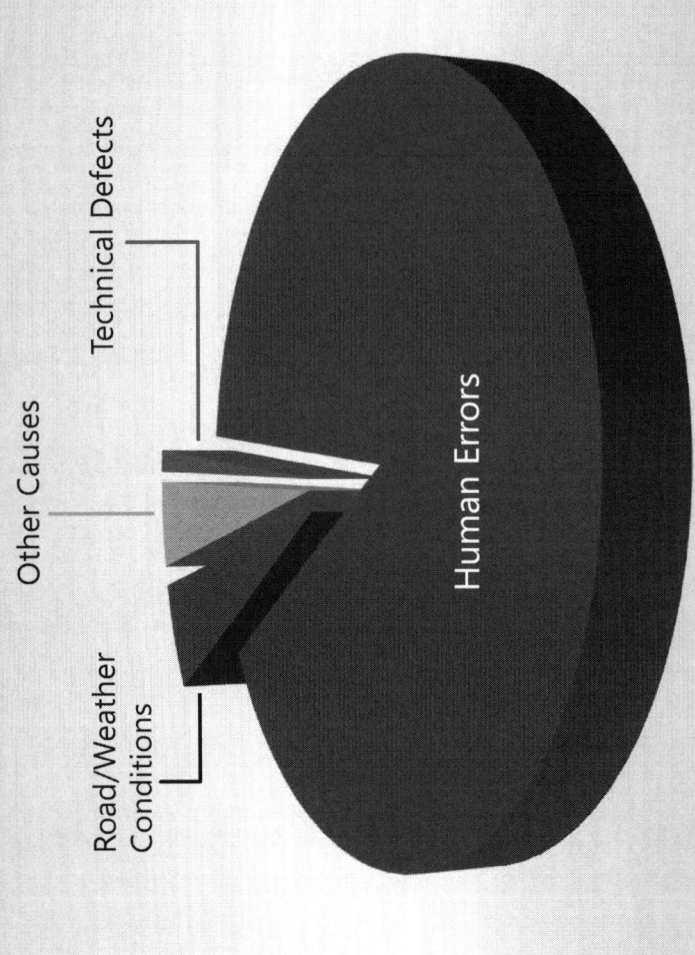

Main Causes of Accidents

Technical Defects

Other Causes

Road/Weather Conditions

Human Errors

Euro NCAP AEB Tests: City and InterUrban

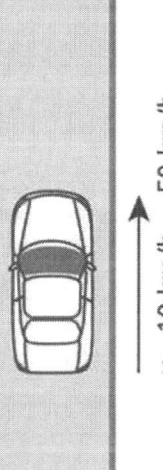
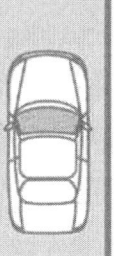
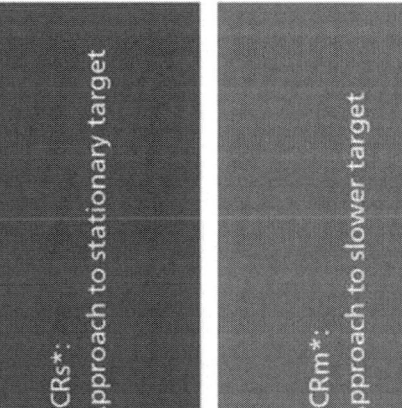
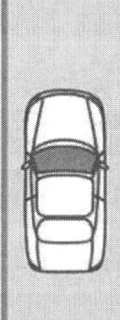
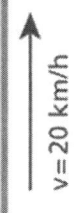
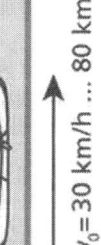

CCRs*:
Approach to stationary target

$v_0 = 10$ km/h ... 50 km/h
$v_0 = 30$ km/h ... 80 km/h

$v = 0$ km/h

CCRm*:
Approach to slower target

$v_0 = 30$ km/h ... 80 km/h

$v = 20$ km/h

CCRb*:
Approach to braking target

$v_0 = 50$ km/h
$v_0 = 50$ km/h
$v_0 = 50$ km/h
$v_0 = 50$ km/h

$d_0 = 12$ m
$d_0 = 40$ m
$d_0 = 12$ m
$d_0 = 40$ m

$v_0 = 50$ km/h, $a = -2$ m/s^2
$v_0 = 50$ km/h, $a = -2$ m/s^2
$v_0 = 50$ km/h, $a = -6$ m/s^2
$v_0 = 50$ km/h, $a = -6$ m/s^2

d_0

* CCR: Car-To-Car Rear;
s: standing; m: moving; b: braking

Embedded Success   **dSPACE**

Euro NCAP AEB Tests: AEB Pedestrian

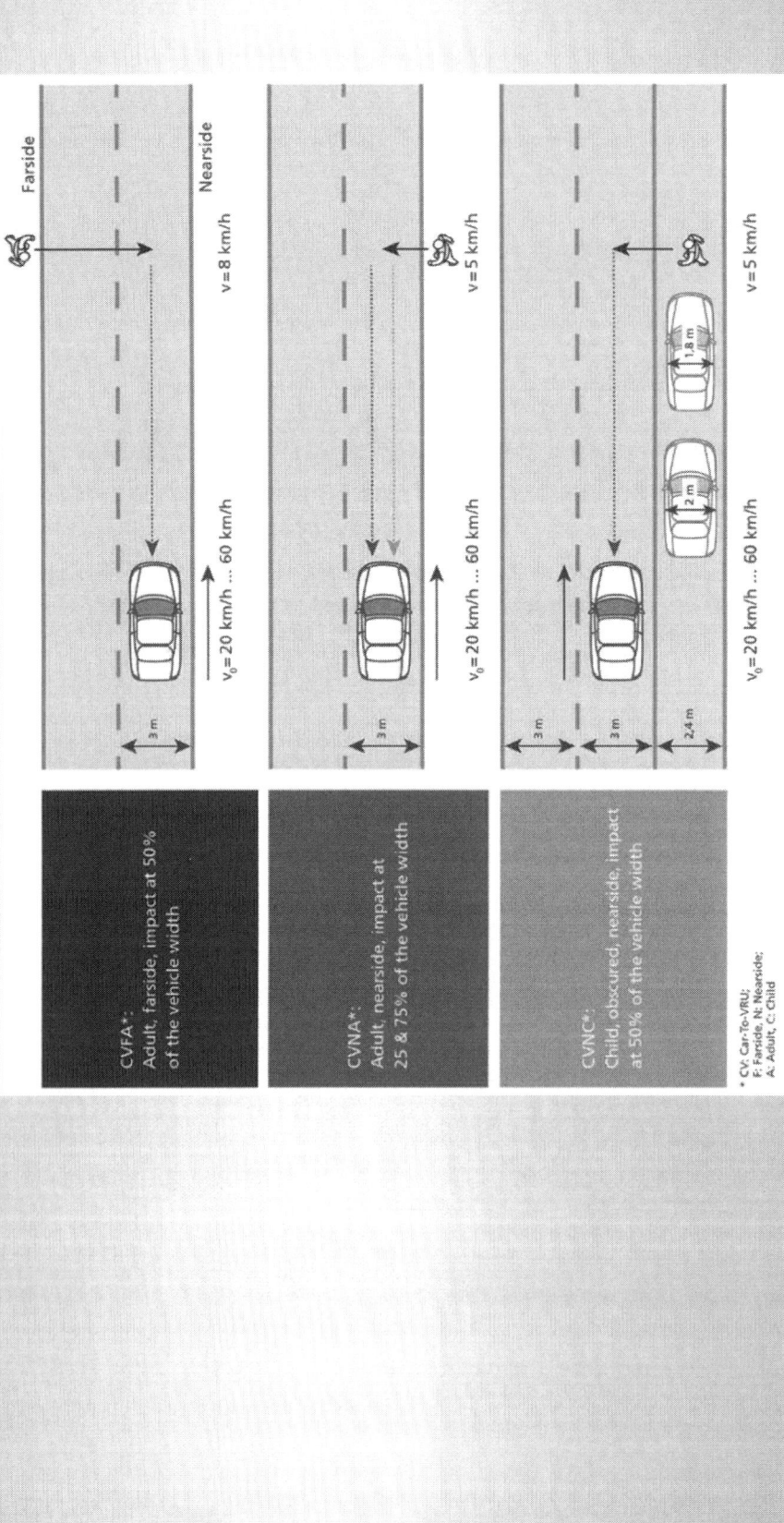

CVFA*:
Adult, farside, impact at 50% of the vehicle width

CVNA*:
Adult, nearside, impact at 25 & 75% of the vehicle width

CVNC*:
Child, obscured, nearside, impact at 50% of the vehicle width

* CV: Car-To-VRU;
F: Farside, N: Nearside;
A: Adult, C: Child

Farside

Nearside

$v = 8$ km/h

$v_0 = 20$ km/h ... 60 km/h

3 m

$v = 5$ km/h

$v_0 = 20$ km/h ... 60 km/h

3 m

1,8 m

2 m

$v = 5$ km/h

$v_0 = 20$ km/h ... 60 km/h

3 m

3 m

2,4 m

Testing Safety-Critical Assistance Systems

- Intensive testing required to ensure functional safety und robustness

- Real-world tests not sufficient, due to system complexity and variety of traffic situations

- Quality assurance according to ISO 26262 required for safety-critical systems

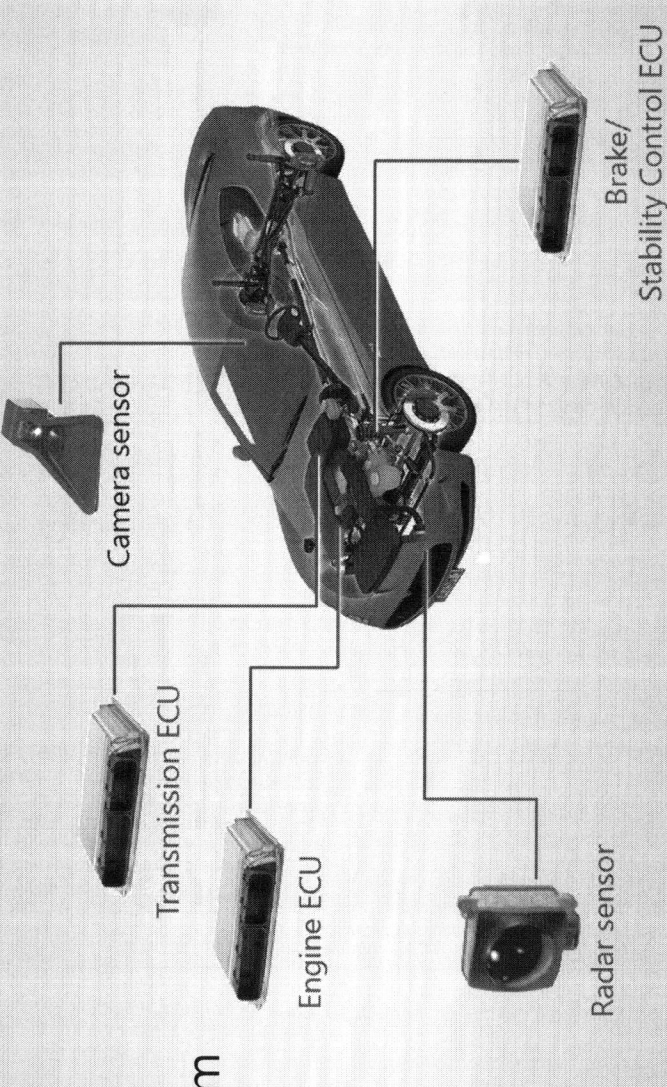

Camera sensor

Transmission ECU

Engine ECU

Radar sensor

Brake/
Stability Control ECU

ADAS and ISO 26262

- ADAS systems influence steering, braking, acceleration – erroneous intervention can be safety-critical!

- Safety-critical scenarios
  - Erroneous Autonomous Emergency Braking
  - Adaptive Cruise Control at full speed

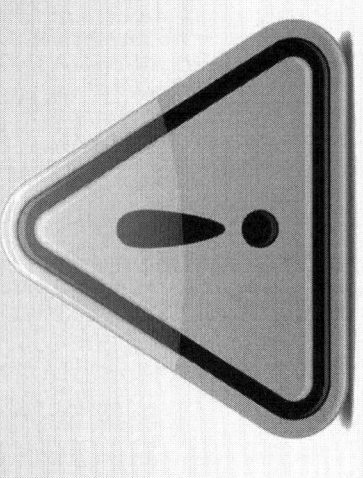

Validate Vehicle Behavior by means of Simulation

Model-in-the-loop test (MIL) → Software-in-the-loop test (SIL) → Hardware-in-the-loop test (HIL) → Real-world test

- ISO 26262 recommends MIL/SIL/HIL simulation for conducting the software safety requirements verification

- UN ECE 13 allows simulation for approval of vehicles with regard to braking

Embedded Success   **dSPACE**

7

One Tool Chain for MIL/SIL/HIL Testing

ISO 26262 ready. Prequalified for all ASILs

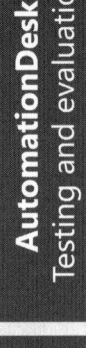

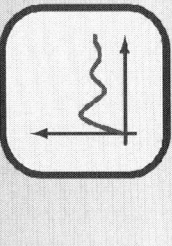

SYNECT
Data management

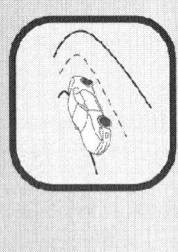

ASM
Open simulation models

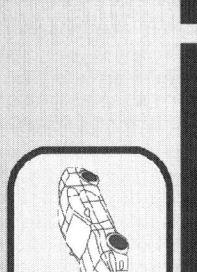
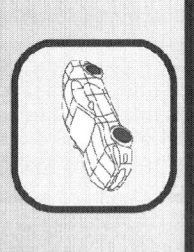

MotionDesk
3-D visualization

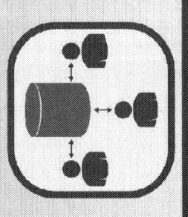

ControlDesk
Experiment environment

AutomationDesk
Testing and evaluation

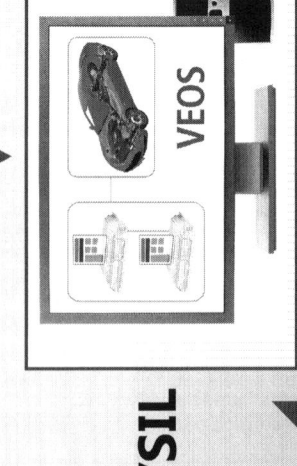

HIL

Real-time validation of components and system

VEOS

MIL/SIL

Early PC-based validation of ECU software and functions

Seamless reuse of data

ASM: dSPACE Automotive Simulation Models

Embedded Success **dSPACE**

8

Test Management for ADAS: Keeping Track of Testing Activities

- **Traceability** and **Coverage**
  - Traceability from requirement to test result and overall requirements coverage
  - For all types of requirements, e.g. safety, functional, performance or robustness requirements

- **Test Scenario Traceability**
  - Which test scenario is tested by which tests?
  - Has a given test scenario been tested successfully?

- Monitor progress across **multiple test platforms** and **different test tools**
  - For tests by means of simulation (MIL/SIL/HIL) as well as real-world tests
  - Test reports, results overviews and test evaluation

- **Test Stimuli Traceability**
  - Which parameters and inputs were used for which tests during a test execution?

Embedded Success   **dSPACE**

9

Test Management for ADAS: Requirements Traceability and Coverage

- All relevant information in one place: directly associate requirements and test specifications with test activities
  - Monitor the requirements coverage progress
  - Identify gaps in the requirements coverage

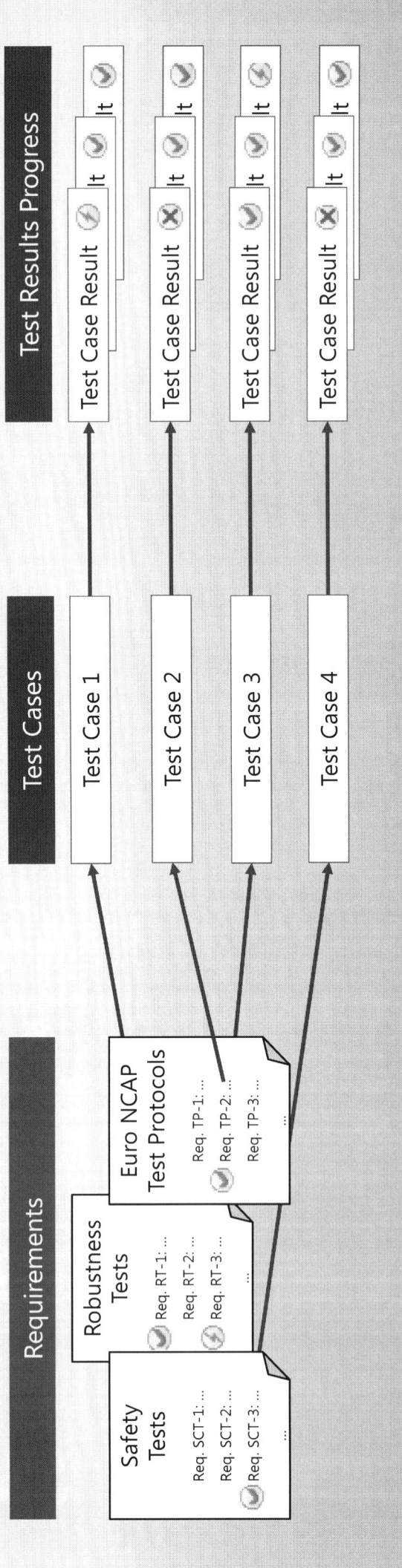

Test Management for ADAS: Test Scenario Traceability

- Manage all test relevant information including tested scenarios
- Identify which test scenarios have been used for which test cases
- Monitor the state of all tests for a given test scenario

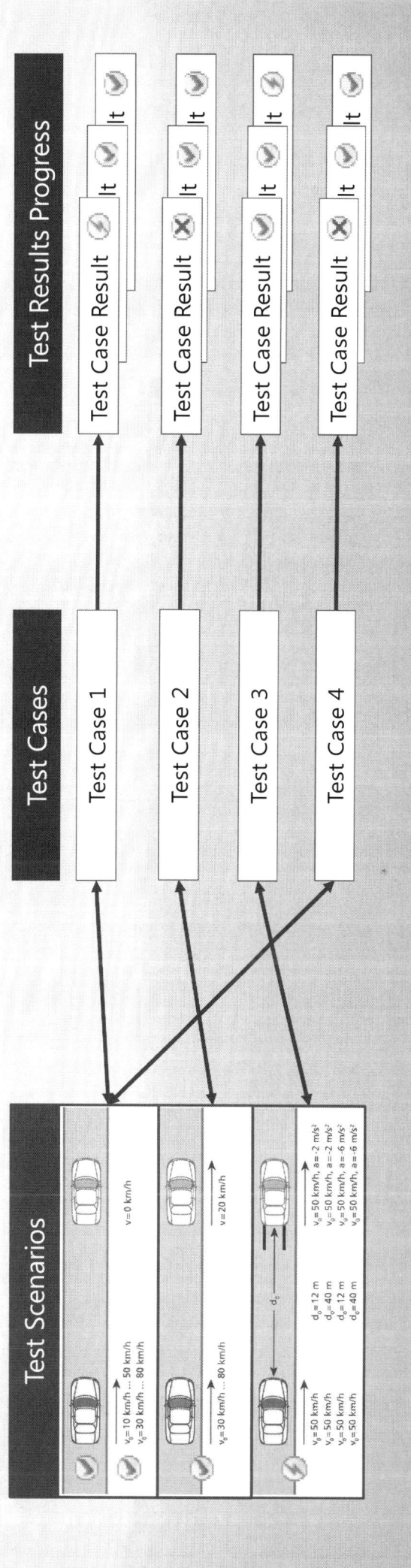

Embedded Success dSPACE

Test Management for ADAS: Euro NCAP AEB Use Case

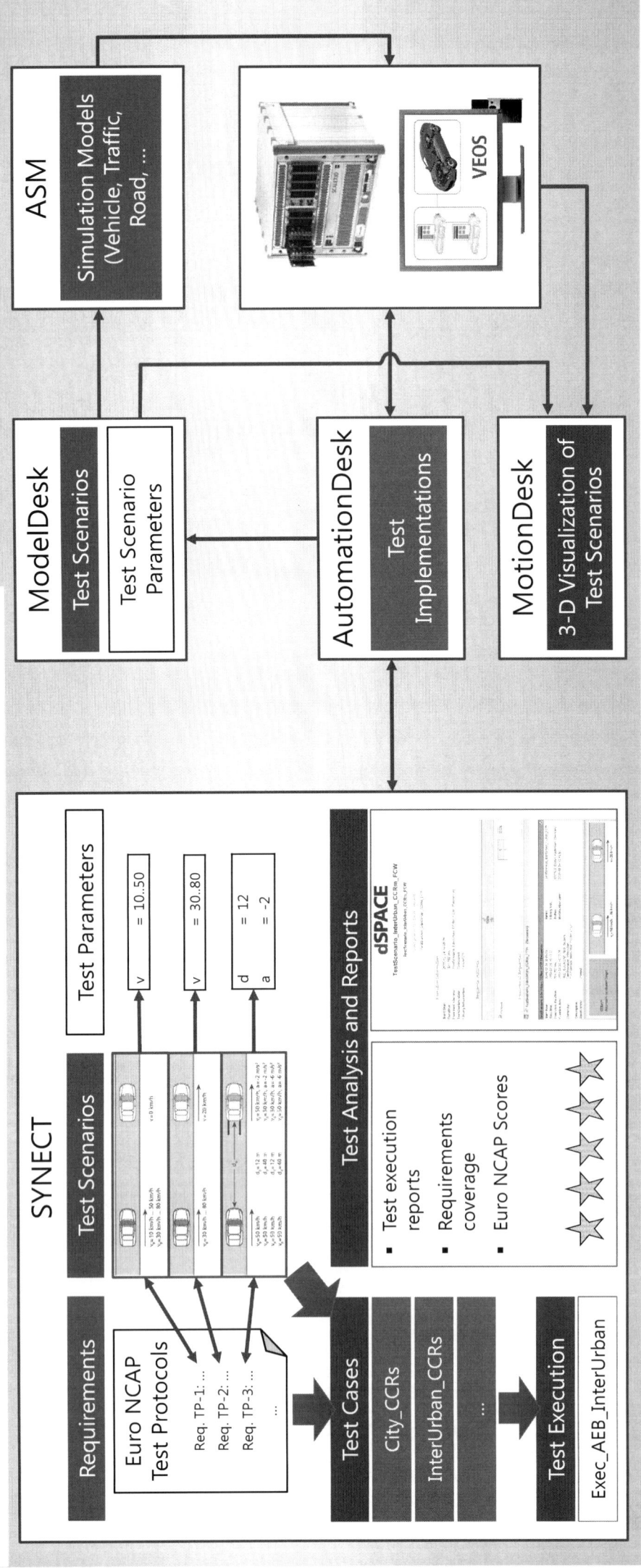

12

Euro NCAP AEB Use Case: Requirements and Test Scenario Traceability

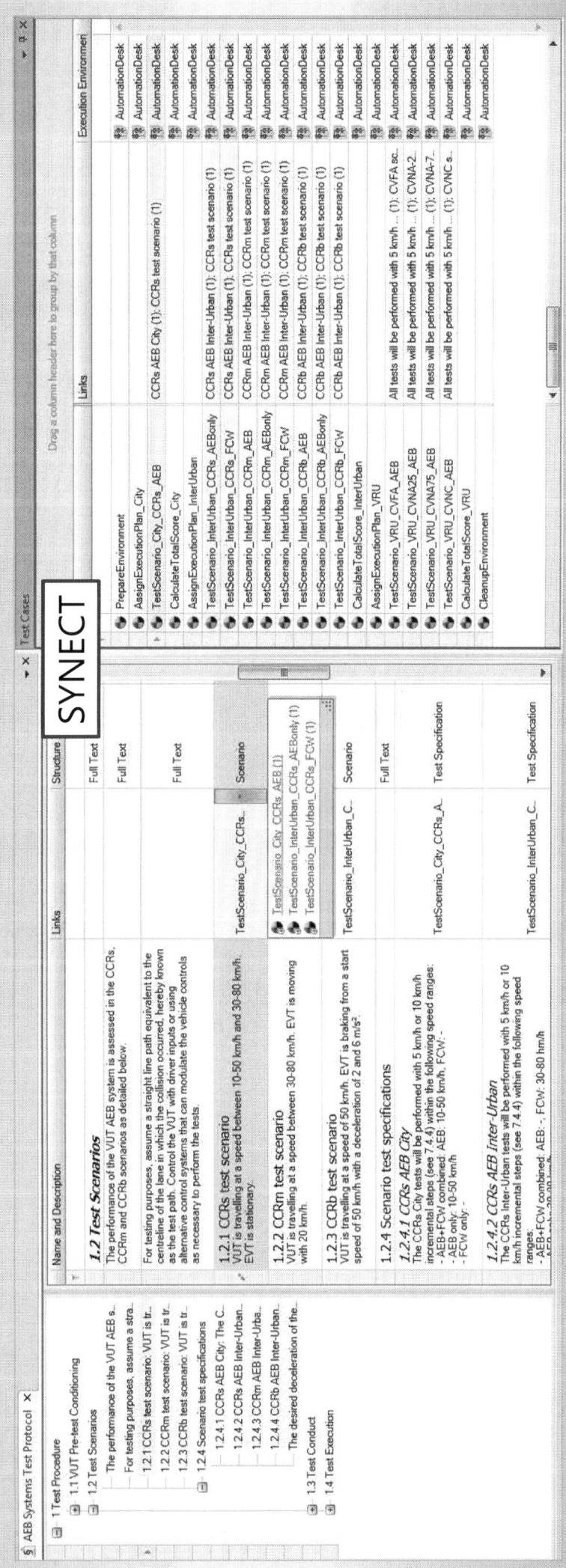

Euro NCAP Test Protocol document in SYNECT with test scenarios linked to managed test cases

Embedded Success **dSPACE**

Euro NCAP AEB Use Case: Planning Test Executions

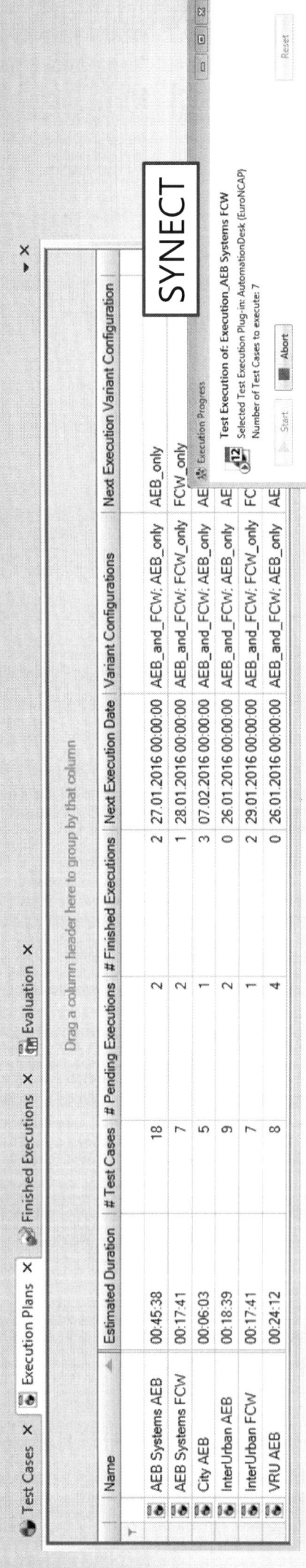

Euro NCAP test execution plans for testing the Autonomous Emergency Braking (AEB) or Forward Collision Warning (FCW) safety systems: execution plans for testing all AEB systems scenarios or for testing specifically City, InterUrban or VRU

Progress of a running test execution in SYNECT

Embedded Success   **dSPACE**

Euro NCAP AEB Use Case: Simulating Euro NCAP using the dSPACE ASM Traffic Model

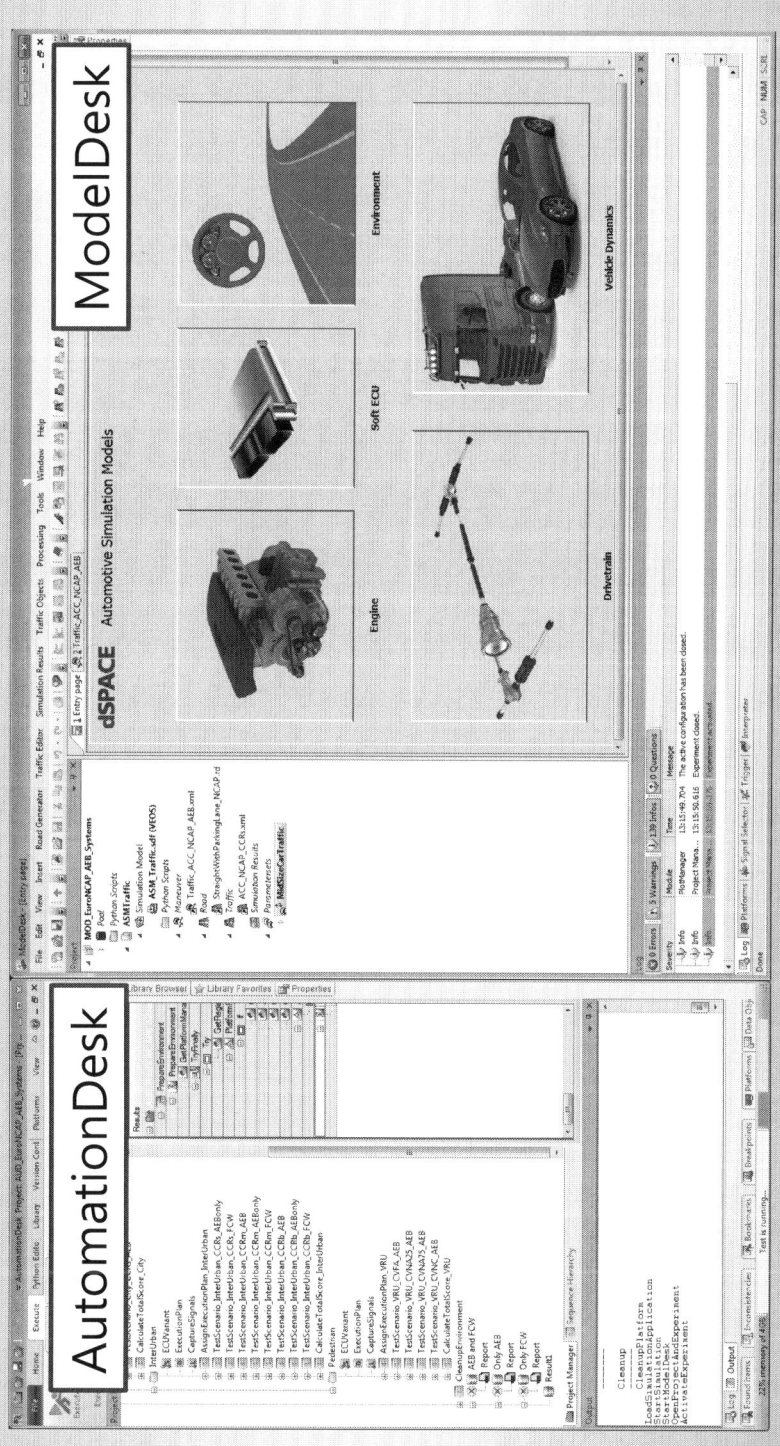

Running test execution in AutomationDesk while parameterizing the test scenario using ModelDesk

Embedded Success dSPACE

15

Euro NCAP AEB Use Case: Visualizing the Simulation

MotionDesk

Visualization of a Euro NCAP Car-to-VRU Nearside Child (CVNC) test scenario simulation

Embedded Success **dSPACE**

Euro NCAP AEB Use Case: Determining the Euro NCAP Rating

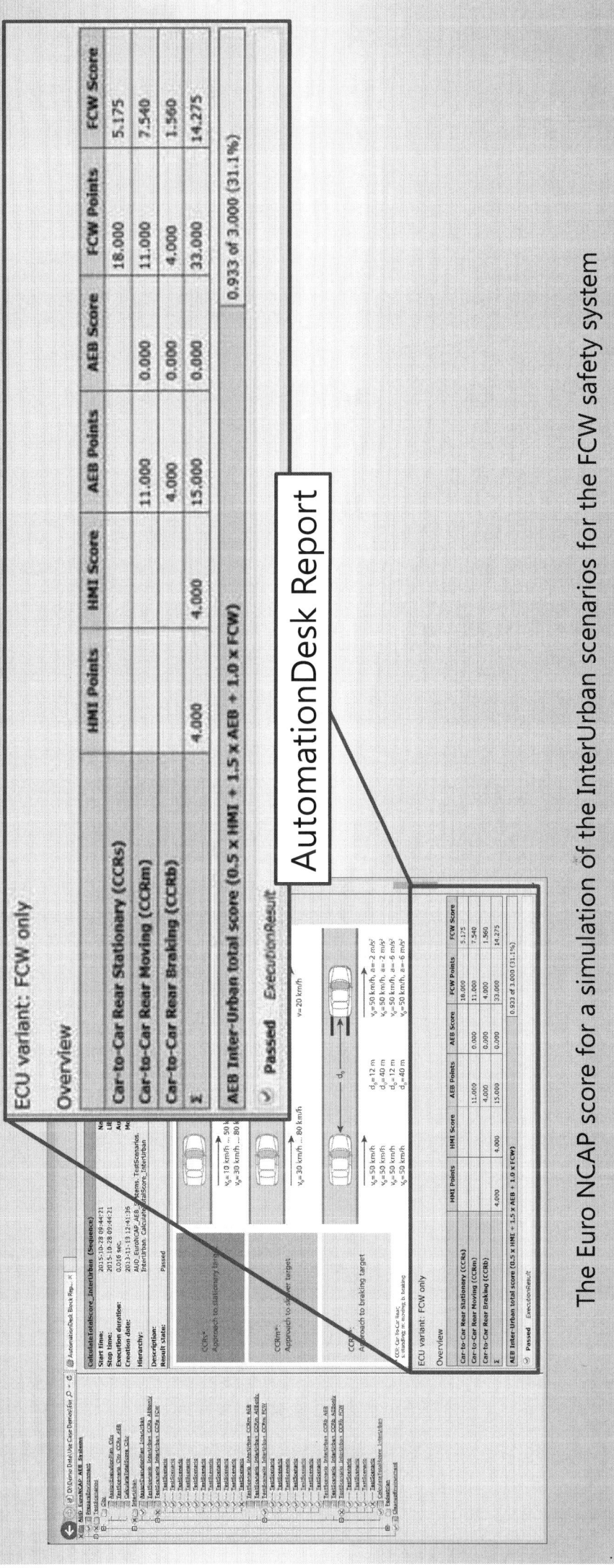

AutomationDesk Report

The Euro NCAP score for a simulation of the InterUrban scenarios for the FCW safety system

Embedded Success **dSPACE**

Euro NCAP AEB Use Case: Requirements Coverage

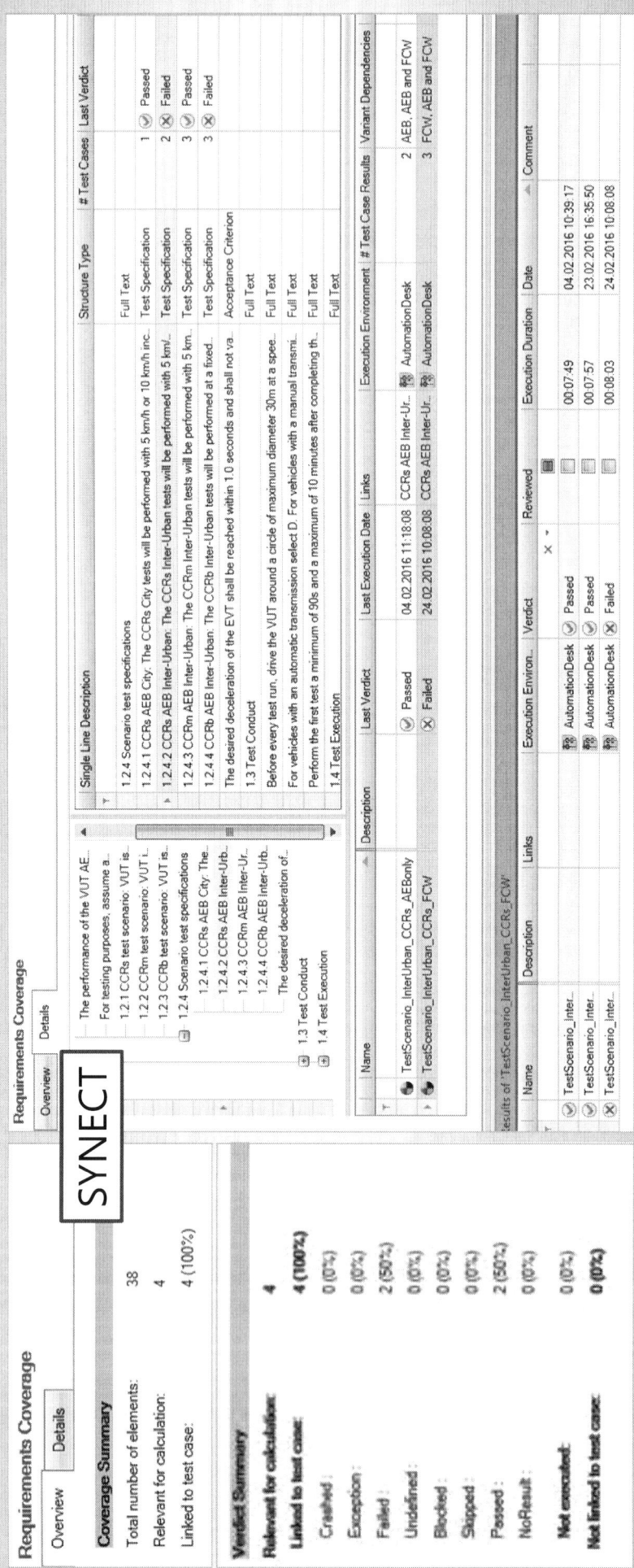

Requirements Coverage of the Euro NCAP Test Protocol document in SYNECT

Embedded Success **dSPACE**

18

Summary

- Advanced driver assistance systems are increasingly important and safety-critical according to ISO 26262

- Testing by means of simulation extends test coverage and allows the robustness of the system to be evaluated

- ISO 26262
  - requires systematic verification of safety requirements
  - requires full traceability from requirements to test cases
  - recommends simulation for verification

- dSPACE provides an integrated tool chain for testing safety-critical assistance systems

Embedded Success **dSPACE**

APPENDIX

ISO 26262 References

- Specification and management of safety requirements (Part 8, chapter 6 (8-6))

- Testing of safety requirements needs to be planned, specified, executed, evaluated and documented in a systematic manner (8-9.2)

- Unambiguous statement of whether the verification passed or failed (8-9.4.3.2)

- MIL, SIL, HIL testing and simulation recommended for Software Unit Testing (6-9), Software Integration Testing (6-10), Verification of Software Safety Requirements (6-11) and Item Integration and Testing (4-8)

Embedded Success **dSPACE**

UN ECE Regulations No. 13 for approval of vehicles with regard to braking

Statements out of the regulation:

- "..... **The vehicle stability function shall be demonstrated** to the Technical Service by dynamic maneuvers **on one vehicle.** This may be realized by a comparison of results obtained with the vehicle stability function enabled and disabled for a given load condition."

- "..... As an alternative to carrying-out dynamic maneuvers **for other vehicles and other load conditions,** fitted with the same vehicle stability system, the results from actual vehicle tests or **computer simulations may be submitted."**

dSPACE Automotive Simulation Models

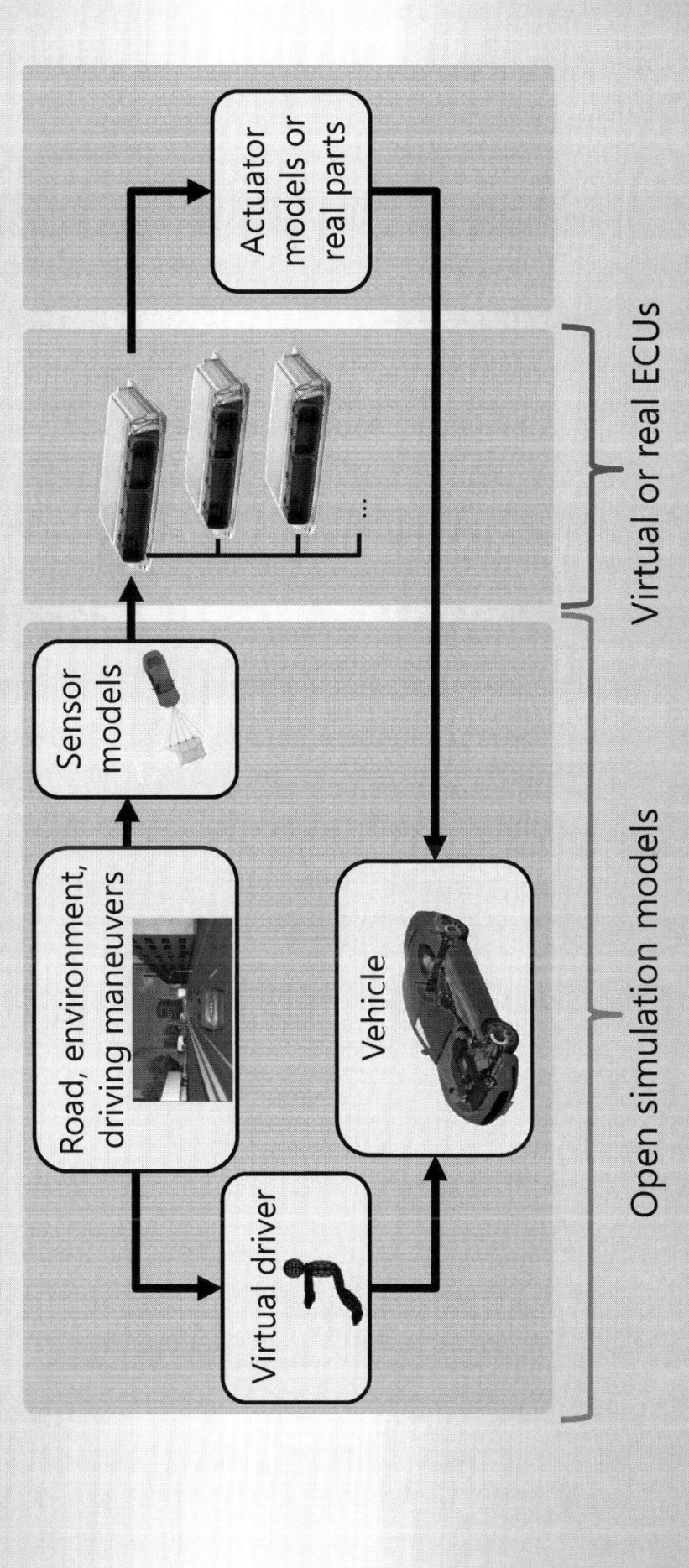

Road, environment, driving maneuvers

Sensor models

Virtual driver

Vehicle

Actuator models or real parts

Virtual or real ECUs

Open simulation models

Test Traceability Across Multiple Test Platforms and for Different Test Tools

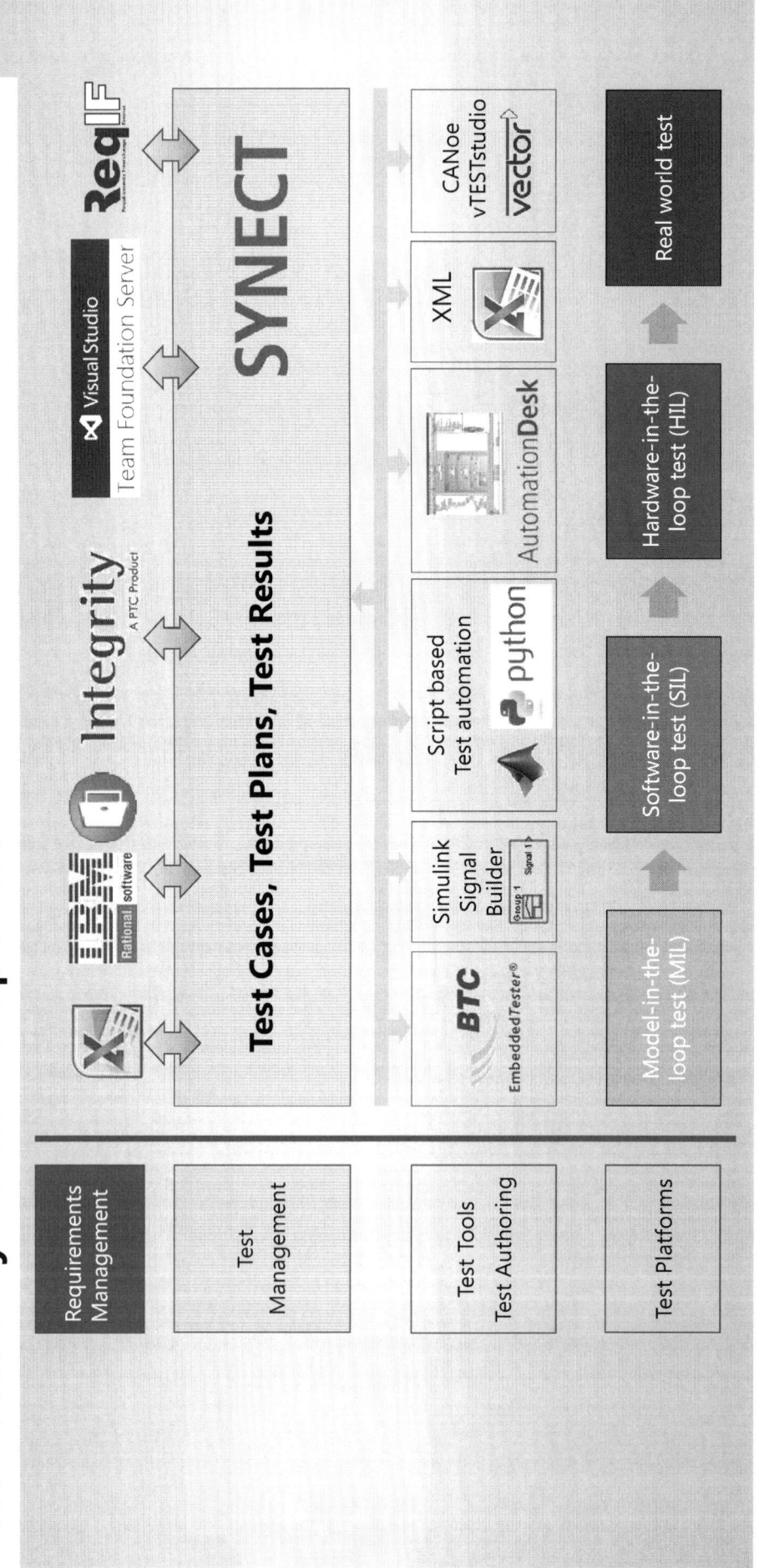

dSPACE

Embedded Success

Embedded Success **dSPACE**

Automated scenario generation for testing advanced driver assistance systems based on post-processed reference laser scanner data

Andreas Wagener M.Sc., IPG Automotive GmbH
Dr Roman Katz, Ibeo Automotive Systems GmbH

1 Abstract

Efficient testing of advanced driver assistance systems is a crucial element for guaranteeing the safe operation of such systems on the road. This paper presents the use of advanced perception systems for automated generation of simulated driving scenarios. The authors describe their progress in the fields of laser scanning processing for reference generation, for both relevant users and road information, and present the entire pipeline for constructing simulated virtual scenarios from real data that can be imported, manipulated and used within a virtual environment.

2 Introduction

The development of advanced driver assistance systems (ADAS) in recent years has taken place at a rapid pace. Existing systems are undergoing further development, and many new ones are being added. As they provide the basis for highly or fully automated driving, major efforts are made to drive their development forward. Forecasts by several automotive OEMs and suppliers, such as Bosch [1] are predicting that highly automated driving will have become reality by as early as 2020. The same can be read in Continental's roadmap [2]. In practical terms this means: automated driving will initially be possible only on motorways and the driver will have to be able to assume control of the vehicle at any time. However, fully automated driving or autonomous vehicles will be following in the foreseeable future – therefore, research and development teams are intensively working on new systems.

In addition to a large number of convenience systems, such as parking assists, many ADAS are safety-relevant. They are intended to react earlier and better than the human driver and help to make 'Vision Zero', i.e. zero traffic fatalities, reality. Their number and level of interlinking has been heavily increasing, which poses new challenges to vehicle developers, as the development process has fundamentally changed: the system can no longer be developed and tested in isolation but, for one, has to be validated in consideration of its interaction with other systems and, for the other, in the context of the whole vehicle including the vehicle's environment. In contrast to conventional testing methods, this involves many other road users. There are various approaches to accomplishing the test of ADAS scenarios. Among other things, they include real-world test setups that operate with pedestrian dummies, for example. This is feasible for defined, limited scenarios but represents only a partial aspect of all the scenarios to be tested in the advanced driver assistance environment. However, success can only be achieved if an entire environment is reproducible – and this is only possible by means of virtualisation. Nevertheless, at the end of a development process, the real-world road test is always the final validation step, but the functionality of the sensor technologies must be ensured much earlier.

Advanced driver assistance systems are currently based on various sensor technologies and combine advanced algorithms for sensor fusion, object tracking, classification, risk estimation and vehicle control, which makes the testing of such systems even more complex. As a result, it has become unfeasible to check the performance of advanced driver assistance systems in the traditional way by driving around, storing data, manually labelling the data for reference, and manually evaluating the results. It is therefore necessary to develop smart and comprehensive reference systems and approaches to cope with the testing of these systems.

The approach presented in this paper addresses these difficulties by installing an additional Reference Sensor System (RSS) on the vehicle, which is already equipped with the prototype (sensor) system that is going to be tested (Device Under Test, or DUT). The recorded data from the RSS is processed to automatically create reference scenarios, including automatic labelling of objects and maps of the roads. Based on this reference data, the DUT can be automatically and objectively verified. The reference data from the RSS can also be converted into a set of simulated scenarios which can be used within a CAE environment. These simulated ground-truth reference scenarios that can be loaded and manipulated provide engineers with a useful platform for checking the performance or consequences of design changes or new concepts.

3 Motivation

Some systems, for instance, are reused in various model ranges, which means that the scenarios which occurred while test driving a particular vehicle model may well be relevant to the development of another one and provide a good data set. This will be increasing even further as more and more new vehicles are being fitted with advanced driver assistance systems. Figure 1 shows the percentages of assistance functions for Germany in 2013 [4]. Emergency brake assists, for instance, can be found in every tenth new vehicle – the enormous development effort resulting from the implementation of this system is easy to imagine.

In systems like these not only the road but pedestrians or cyclists must be considered as well. Questions that typically arise are: Will they be correctly detected? At what distance will the system be triggered? What data will be used as the basis? This wide variety of factors results in the requirement of testing the system functions as early as possible, which falls into a stage of the development process in which no real-world prototype exists yet. In order to be able to test the functions in spite of this, reproducible tests in virtual test driving provide a viable approach. They must be based on realistic data obtained in real-world road tests.

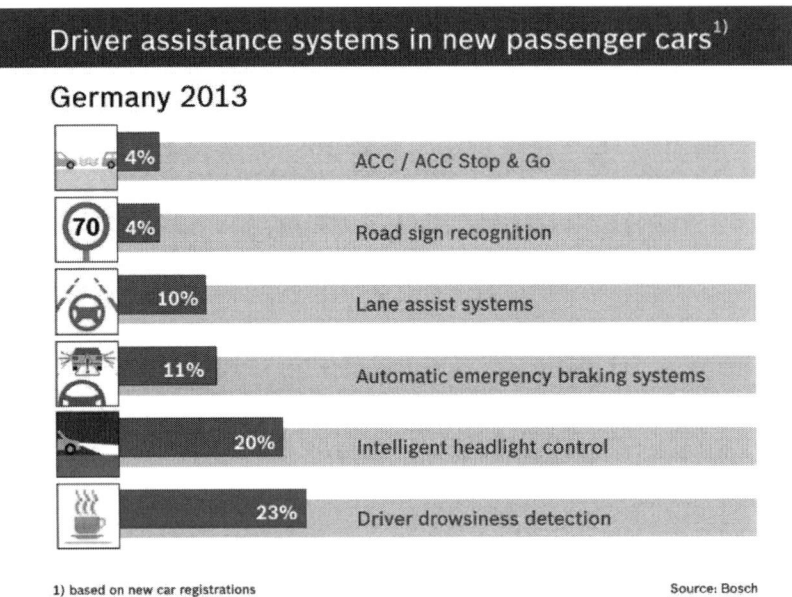

Figure 1: Percentages of the six-top selling advanced driver assistance systems in new cars in 2013

4 Testing of ADAS Scenarios in Virtual Test Driving

4.1 Relevance of the Environment for ADAS Testing

One of the problems in validating advanced driver assistance systems is a high dependency on the environment. Scenarios that provoke a critical condition, a malfunction or an undesirable behaviour cannot simply be simulated and repeated. For the development engineer, however, the repetition of a malfunction is important for analysing the noted malfunction in greater detail.

But for real-world road tests it is often impossible to exactly reproduce critical scenarios. For one, they are too complex and for the other, their simulation would either not be economically feasible [5] or could not even be influenced. Relevant examples are situations at construction sites, driving in specific weather conditions (e.g. intensity of rain) or multi-lane intersections, as depicted in Figure 2.

Figure 2: Intersection scenario with various environmental factors
© MACIEJ NOSKOWSKI, iStock

For the system test, however, this entire environment has to be taken into account. The resulting complexity (for concrete factors see Figure 3) for the development engineer consists of both static and dynamic influences, as well as general external conditions. Their number (road users, time of day, etc.) is too large for creating a reproducible scenario. Two examples illustrate the point:

In the case of an emergency brake assist system, for instance, this could be an intersection scenario with a suddenly crossing road user. A scenario like this, without considering the economic element, entails too many influencing factors, such as other vehicles involved, pedestrians or cyclists, to be repeatable with sufficient accuracy.

A function that warns drivers of objects in their blind spot or monitors lane changing is primarily used on motorways. There, speed is high and the number of road users large, and therefore difficult to simulate. At the same time, the risk, due to high speeds and the associated dynamics, is particularly high. Critical for such a function is the recognition of traffic objects of various forms and sizes, as well as speed, in order to estimate the potential of danger.

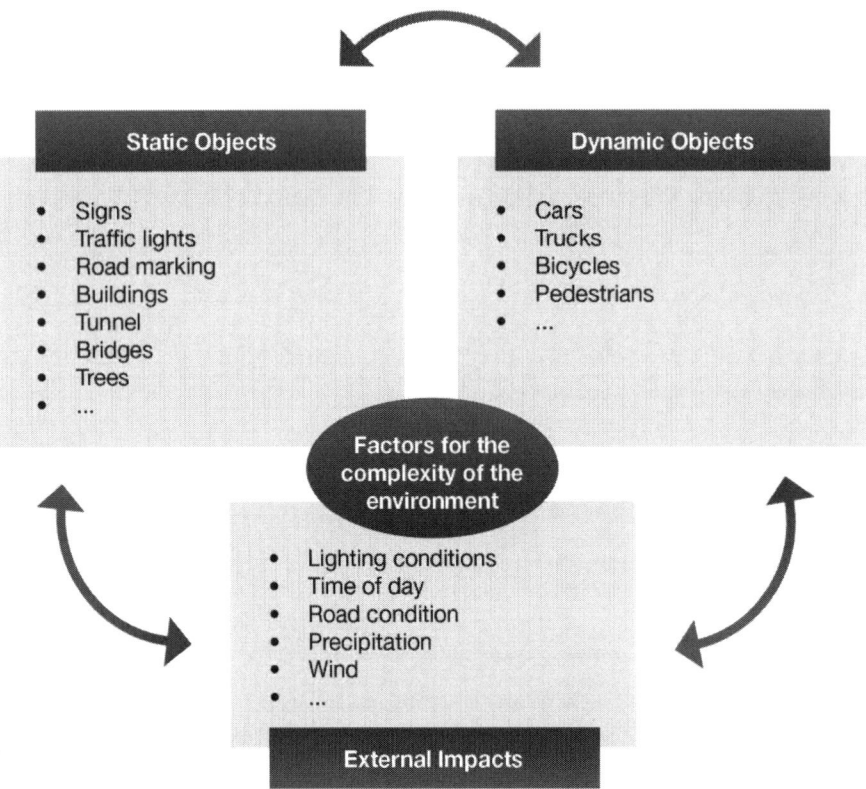

Figure 3: Factors which influence the complexity of the environment

A previous approach has been to model a similar situation in a simulation, which involves a significant modelling effort (manoeuvres must be created for all traffic objects and the selected sequence adjusted). The data acquired by the vehicle sensors (in addition to various radar, lidar and ultrasonic data, this also includes video footage from diverse cameras) provide the reference for the modelled traffic objects. Sensor inaccuracies and sensor faults may distort the scenario or limit the depth of detailing, which may lead to consequential errors in the simulation.

In turn, as the number and complexity of advanced driver assistance systems keeps growing, so does the effort required to validate them. Still, it is not unusual for cases which cause a malfunction of the ADAS only being discovered at a late stage of the product development cycle, typically during the initial real-world road tests. Ideally, these test cases should be reproducibly stored for reuse in the simulation.

4.2   Requirements Imposed on a Solution

In order to be able to validate assistance systems better and more efficiently in future, a method is required that can reproduce critical test cases in test driving. At the same time, storage of the test cases is desirable to enable frontloading for the next development cycle. Additional requirements for the next generation of ADAS, which in the ADAS context are typically safety-critical, can thus be captured with greater accuracy, and investigated and validated at an earlier point in time during the development process.

The V-Model that is popular in the automotive industry for the development of functions has the disadvantage of providing for real-world road tests only at a late stage of the development process. In the development of ADAS functions, due to the high dependency on the environment, road tests generate new scenarios that were not considered in the scope of the requirements analysis at the beginning of the development cycle. These scenarios must be preserved in order to be able to take them into account by means of frontloading in the next development cycle as early as when the requirements are being defined. Ideally, the proposed method is suitable for use in all stages of the development cycle.

4.3   Overview of Reference Technologies at Ibeo

The generation of virtual scenarios from real data focuses on the use of specific technologies for generating reference data. These technologies are part of a set of enablers that have been developed by Ibeo Automotive Systems. The implemented taxonomy includes not only the generation of ground-truth reference data from real-world driving, but also the specific implementation of ADAS/HAD, and automation and tooling functionalities. Figure 4 shows the classification of enabling technologies for testing ADAS/HAD as pursued by Ibeo.

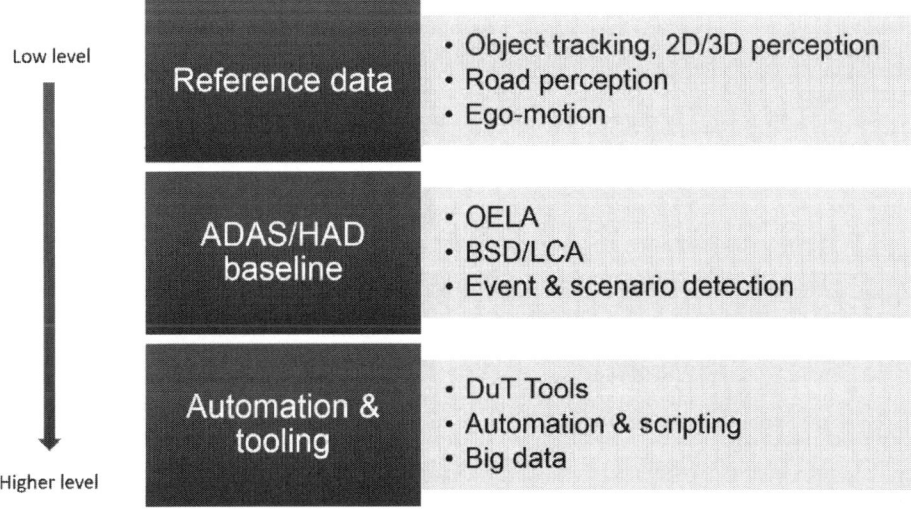

Figure 4: Enabling technologies for testing ADAS/HAD as pursued by Ibeo Automotive Systems

A comprehensive overview of testing approaches and an alternative taxonomy can also be found in [6]. At the core of this hierarchy lies the reference data which provides the ground-truth data from real-world driving and provides the basis which all other technologies for testing ADAS/HAD build upon. This level includes the perception of objects (2D and 3D-based), the perception of infrastructure and road information, and the inference of accurate ego-motion data.

The ADAS/HAD baseline level builds on the reference data level by implementing reference ADAS/HAD applications such as BSD/LCA or Object-to-ego lane association (OELA). Automation and tooling constitutes the higher level of abstraction in this hierarchy providing, for instance, functionalities to efficiently process data and DUT tools.

5   Details of the Approach

This section presents the technologies used to obtain data collection from a test vehicle, to obtain reference objects and reference road information from this collected data, and explains how the obtained reference data can be imported into the simulator.

5.1 Data Collection

The data collection involves logging raw sensor data that can be used offline to recreate the observed scenes. This typically includes horizontal laser scanners for object detection and down-looking laser scanners for road perception. In addition to this external information, the motion of the test vehicle is logged. Figure 5 illustrates the reference sensor setup normally used for the reference data collection process.

Figure 5: Horizontal laser scanners (left) are used for the perception of objects and relevant road users. A down-looking laser scanner mounted on the roof of the test vehicle (right) is used for road perception.

5.2 Reference Data Generation

Reference Objects

Several advancements in the field of laser scanning allow Ibeo sensor technology [7] to be used as RSS for automatically generating reference scenarios from laser scan data. Current generations of laser scanners can detect objects with very high precision relative to the scanner. If a digital map is available which contains landmarks with centimetre accuracy, then the absolute position of the laser scanner can be determined within a range of a few centimetres, thus improving the accuracy of all tracked objects. The scanner is effectively used as a differential sensor, analogous to Differential GPS (DGPS), but without the need for a base station or for a clear view of the sky.

A further improvement is the use of Ibeo's Evaluation Suite software for offline processing to fully analyse the scenario [8]. State-of-the-art online tracking algorithms are inherently limited by real-time requirements. If the data is reprocessed offline, it becomes possible to look ahead for all observations of an object and associate them with the first instant that the object is visible, which allows the system to be used as a true reference, since no online sensor can look into the future. This also allows the use of a 'best-situation-classifier'. The point in time when an object is most clearly visible can be then used to classify the object, and this classification can be extended to the lifetime of the object. Additional benefits are improved robustness to occlusions, reduced uncertainty of the ego vehicle position between landmarks, and increased accuracy of object trajectories.

Figure 6 introduces the forward-backward tracking used by Ibeo to obtain object reference data. In the online approach (left), the approaching vehicle is initially detected, but it is not yet classified and the outline is not clear. As the vehicle gets closer, the outline can be clearly seen and it is classified as a car. In the offline approach (right), this information is associated with the object in the first instant that it is visible, allowing a more accurate reconstruction of the scenario.

Reference Road Information

Down-looking laser scanners are used to provide very detailed 3D information about the profile of the road including reflectance information. This data is processed offline using Evaluation Suite to automatically detect road markings, guardrails and other valuable road infrastructure and to generate maps of the roads. This technology and some of its features are illustrated in Figure 7.

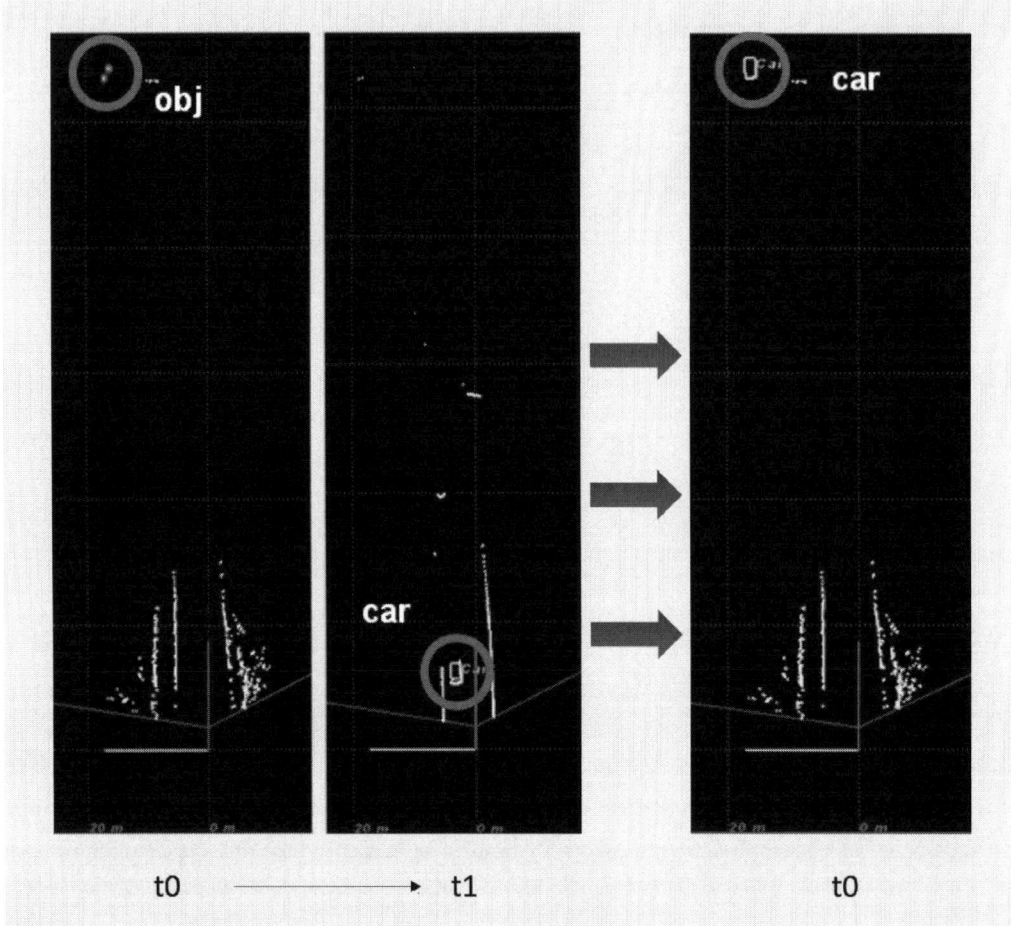

Figure 6: Online (left) and offline (right) results as provided by the forward-backward tracking algorithms used to generate reference data for objects

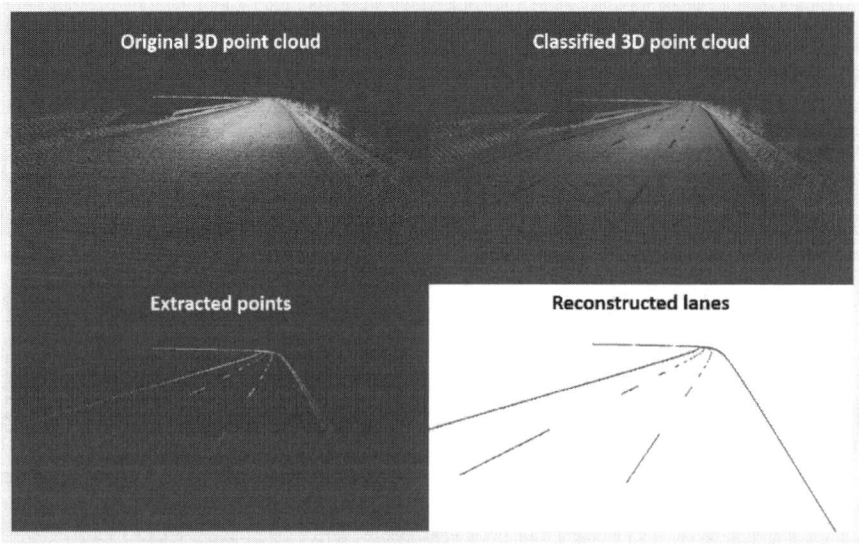

Figure 7: Example of reference road marking maps

The example in Figure 8 illustrates the application of Evaluation Suite for generating reference scenarios that contain reference objects and reference road information. The top image shows the online results, which are improved in the offline stage using Evaluation Suite. The bottom image shows that all the labelled objects have the correct classification results, and their bounding boxes are consistent with the correct underlying class. Moreover, a map of lane markings has been built, and the labelled objects are now provided in combination with this map as the final output.

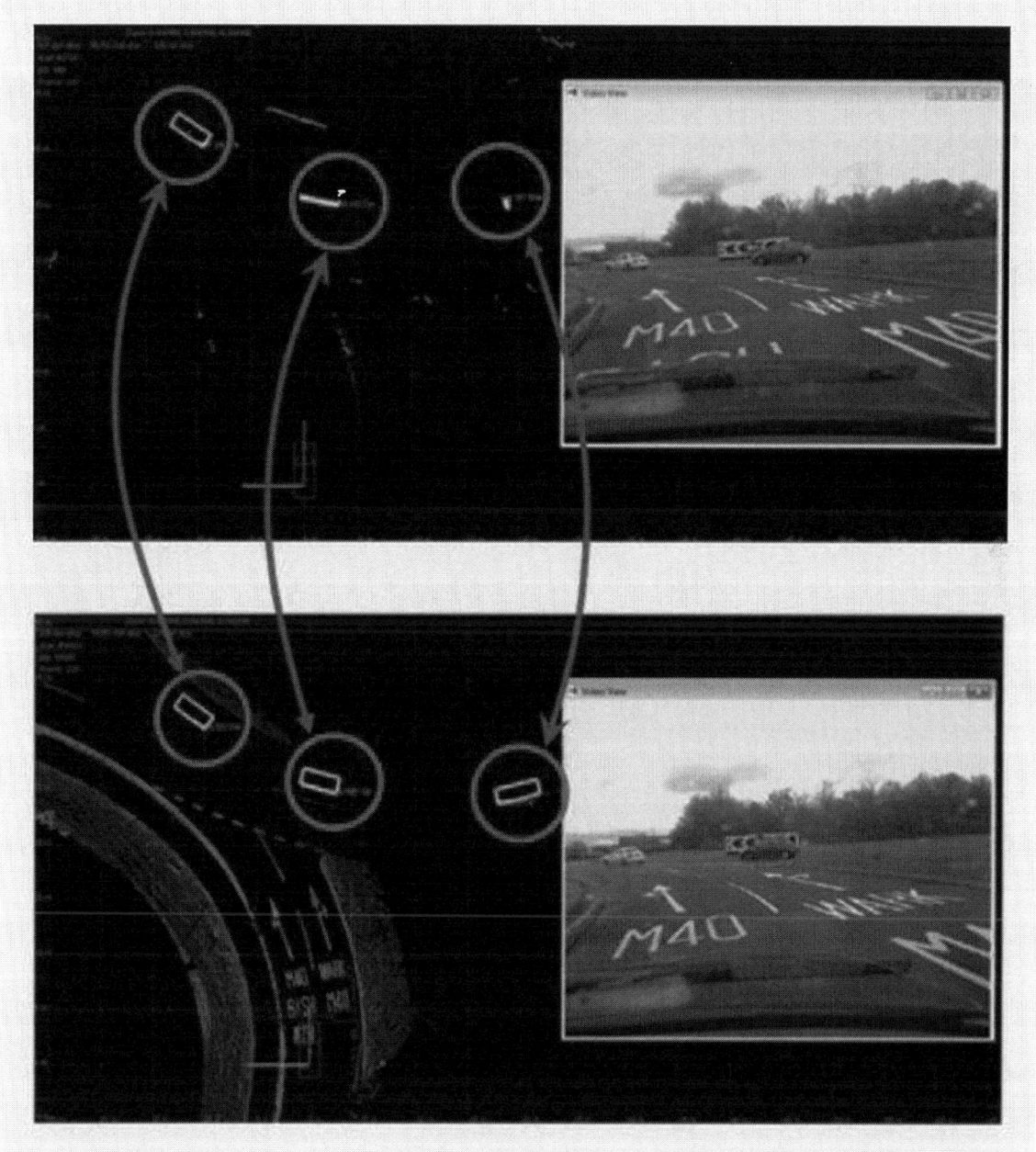

Figure 8: Online (top) and offline (bottom) results as generated by Ibeo's Evaluation Suite

6 Presentation of Data Utilisation in the Simulation

By using laser scanners and a simulation environment for test driving, the reproduction of a real-world scenario can be transferred into the virtual world. For this purpose, the method presented uses post-processed reference laser scanner data.

The workflow of the method is depicted in Figure 9. The method's starting point is a prototype equipped with reference laser scanners in which the system or systems to be tested has or have been installed. If during the road test the test driver notices a faulty behaviour of one of the systems under test, he can save the scenario in a file or database.

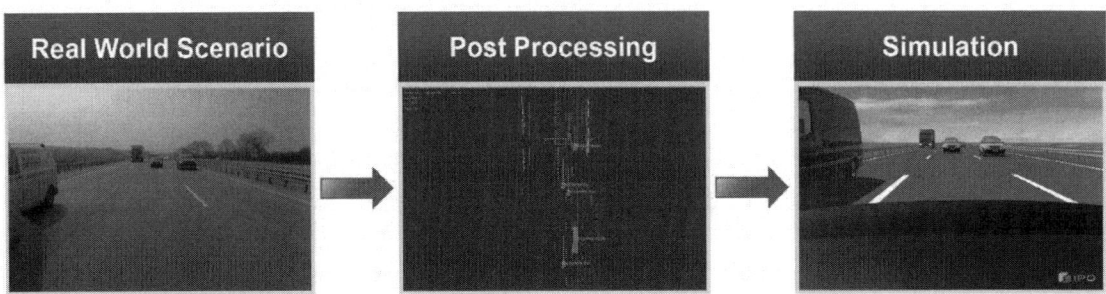

Figure 9: Workflow of the method

Following the processing of the algorithm by Ibeo, the required data can be imported into the CarMaker open integration and test platform. Of particular interest for testing ADAS functions is the environment in the form of road users and traffic objects. In order to represent all road users identified, the Ibeo SDK was integrated into the CarMaker environment. Due to the integration of the Ibeo SDK, no other tool is required and the import of the road users from a logged and post-processed file can be initiated directly in CarMaker. During the import process all the traffic objects are decoupled from the ego vehicle. Internally, an object list is created which is subsequently reloaded for the simulation. By reloading it into the simulation the scenario experienced by the engineer is transferred into the virtual world. At the same time, the image files, which are also stored in the proprietary Ibeo file format, can be extracted. As they are available in JPEG format, they can be easily converted into an appropriate video format. This results in the possibility to make a visual comparison between the simulation and reality. Not currently possible, but planned for the future, is an additional import of the road and map information (e.g. lane markings, number of lanes, etc.). This would enhance the quality of the reconstructed scenario even further.

This procedure can be consistently used in the development process, as illustrated in Figure 10. The method presented in this paper, in conjunction with Model-in-the-Loop (MIL), gives developers the possibility to test available critical scenarios and to validate their initial design early. If software is already available, the process can also be used within the scope of Software-in-the-Loop (SIL). The same applies to the Hardware-in-the-Loop (HIL) setup in which real-world components can be exposed to the recorded scenario. This has the advantage that real-world hardware is investigated under specific conditions, which can considerably facilitate the fault search at this stage for the respective engineer.

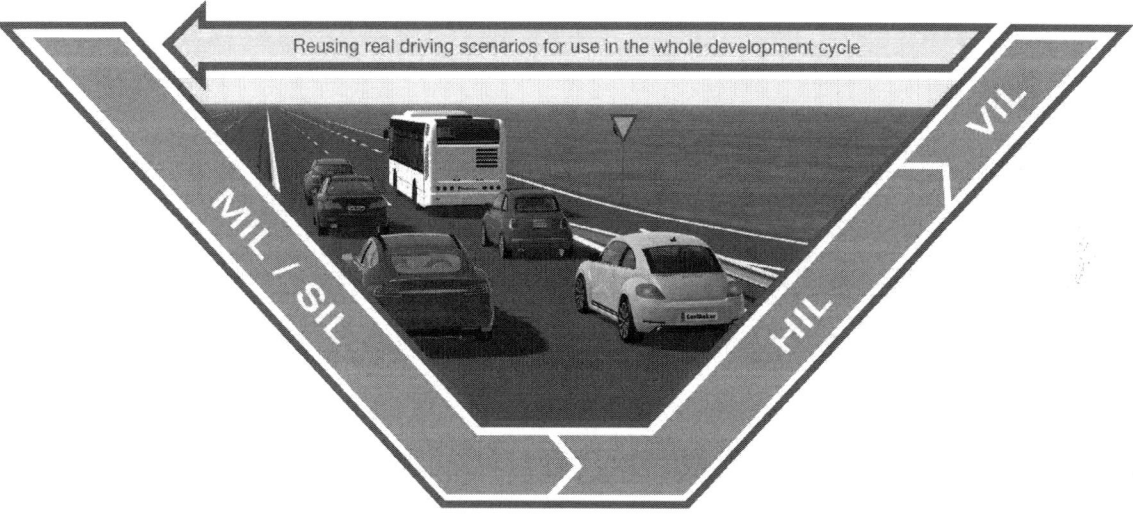

Figure 10: Process for reusing the generated driving scenarios in all stages of a development process

In addition to the fault search, the method, in conjunction with Vehicle-in-the-Loop (VIL), makes it possible to repeat the scenario in a way which can be experienced by the engineer/developer. Particularly with respect to functions for partially and highly automated driving, this is a property not to be neglected. These functions often have subjective elements the requirements of which must be fulfilled. They can be tested by using VIL in combination with the method presented, drawing on real-word traffic scenarios, which means that the behaviour of the environment corresponds to the logged test run. VIL is also advantageous because there is no need for involving real-world vehicles in the test and, for this reason there is no risk of injury to people or damage to vehicles.

7   Experimental Evaluation

This section presents an example that illustrates the processing pipeline. Figure 11 shows reference data for a scenario on the motorway as provided by the Evaluation Suite offline software. The left image shows the reference objects only, whereas the right image includes the reference objects and combines them with reference road information. For comparison, Figure 12 shows the simulated result of the imported scenario from a similar perspective as the camera image in Figure 11. It depicts the identical arrangement of traffic objects within CarMaker.

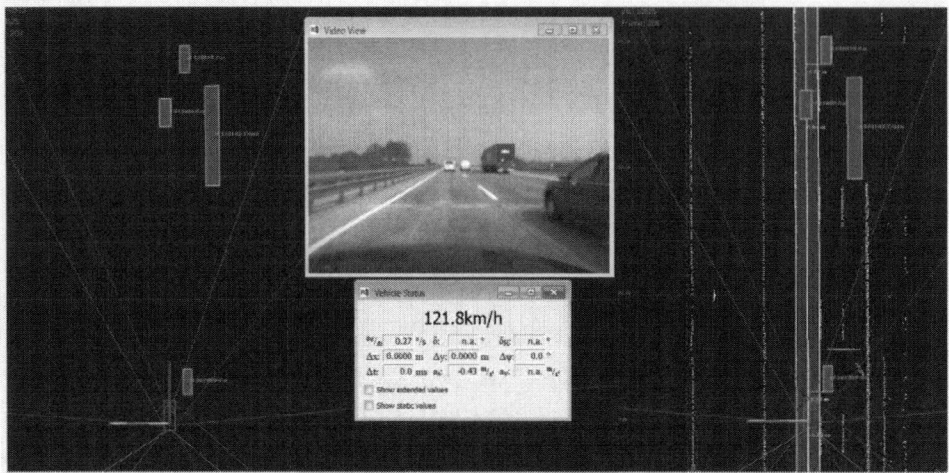

Figure 11: Reference objects (left) and reference objects with lane information (right) as provided by Ibeo's Evaluation Suite offline software

Figure 12: Virtual scenario in CarMaker

8 Conclusions and Outlook

This paper presented an approach for generating realistic scenarios within the CarMaker vehicle simulation environment based on data from real traffic provided by Ibeo's reference sensor systems and post-processing software.

The object reconstruction is general and does not depend on specific vehicle sensors. Moreover, Ibeo's post-processing algorithms are able to gain additional information that increase the quality of the generated scenarios. By importing the object data into the CarMaker simulation environment, a scenario that was driven in the real world can be virtually repeated. In addition to the object data, another laser scanner captures additional environmental information, including static objects such as road markings or road boundaries. The scenarios gained this way can be used for the validation and development of advanced driver assistance systems across the entire development process (from MIL/SIL to HIL and VIL).

Future work includes the import into the simulation environment of road markings and map information. A manipulation of the object list within CarMaker will also be investigated. Such a manipulation is of importance in order to generate a large number of variants from a single scenario. Further investigations to specify the type of required manipulation as the basis for varying the traffic objects are planned. This approach holds major potential to increase efficiency, as the manual setup of a large number of individual scenarios can be omitted. Hence, developers can use a vast array of complex scenarios for functional tests of their assistance function at an early stage and compare the fulfilment of the requirements. Thus, by means of the approach presented, a central issue in the vehicle development process can be resolved: the achievement of high system maturity at an early development stage.

References

[1] Bosch Press Release 18 March 2015. Dr Volkmar Denner at the 14th International Stuttgart Symposium: Bosch putting the autopilot on the road. http://www.bosch-presse.de/presseforum/details.htm?txtID=6730&. Last access: 21 April 2015.

[2] Continental: Mobility of the future: Automated driving. http://www.continental-corporation.com/www/presseportal_com_de/allgemein/automatisiertes-fahren-de/automatisiertes_fahren_intro_de.html. Last access: 21.04.2015.

[3] S. Ulbrich, T. Menzel, A. Reschka, F. Schuldt, M. Maurer: Definition der Begriffe Szene, Situation und Szenario für das automatisierte Fahren, *10. Uni-DAS e.V. Workshop Fahrerassistenzsysteme,* 2015.

[4] Bosch press release 16 September 2014. Bosch evaluation of driver assistance systems: One in every four new cars can detect when drivers are tired. http://www.bosch-presse.de/presseforum/details.htm?txtID=6962&tk_id=108. Last access: 12 February 2016.

[5] S. Hakuli, Stephan; Krug, Markus: Virtuelle Integration. In: Winner et al (eds.): *Handbuch Fahrerassistenzsysteme*, 3. Auflage, 2015. Springer Vieweg, Wiesbaden, Germany. S. 126.

[6] J. Stellet, M. Zofka, M. Zoellner, F. Niewels: Testing methods for advanced driver driver assistance and automated driving. In: Proceedings of the *2nd International VDI Conference*, Düsseldorf, Germany, 2015.

[7] http://www.ibeo-as.com/, Ibeo Automotive Systems GmbH, 2013

[8] M. Spencer, R. Katz, U. Lages: Forward-Backward Object Tracking for Generation of Reference Scenarios Based on Laser Scan Data. In: Proceedings of the *21st ITS World Congress,* Detroit, USA, 2014.

DNA forAutonomous Driving:
Adapting Robot Architectures to AUTOSAR

Dr.-Ing. Björn Giesler, Elektrobit Automotive GmbH

Introduction

Today's driver assistance systems have come a long way from the introduction of the first parking aid and ESC developments in the 1990s [1]. Current-generation vehicles can follow cars, keep and even change lanes, brake automatically, and even become self-driving. But so far, each of these systems often takes care of only one single, rather simple, task: Lane keeping assistants aid by actively steering away from the lane markings, adaptive cruise controls match the vehicle's speed to that of the leading vehicle and keep the proper distance. When the systems are activated at the same time, they act together rather by accident instead of by intention. As the first serious forays are made into semi- or even fully-automated driving, it becomes clear that even in a vehicle that has automated lateral and longitudinal control, coordinating the different aspects of keeping a vehicle on course is not an easy task. One of the main reasons this is so hard is the lack of a proper, scalable systems architecture, which would encompass all driver assistance systems in the vehicle and make it easy to tie them together into a full automation system.

Constraints For Traditional Driver Assistance Systems

There are many constraints that drive development of vehicle systems. Striving for technical excellence is a given, of course; but equally important are physical space[1] the system takes up, the scalability over different vehicle groups, and of course cost. These constraints lead to system design that is packaged in as few "boxes" in the vehicle as possible, and that performs only its designed function and nothing else. An example of

---

[1] and, under increasing pressure to reduce emissions, energy efficiency and weight

such a system can be seen in the highly simplified example of two separate driver assistance systems shown in Picture 1. Here, the "Lane Keeping" is packaged in one box that contains the physical camera itself, as well as the control unit that is responsible for detecting pairs of lane markings and the vehicle's position between them, and determining the correct value for the steering torque controller to help the driver to hold the vehicle in its own lane. Likewise, the "Adaptive Cruise Control" mechanism is packaged in one box that contains the radar sensor proper, plus an ECU with the algorithms running on it. The same goes for the "Automated Emergency Brake".

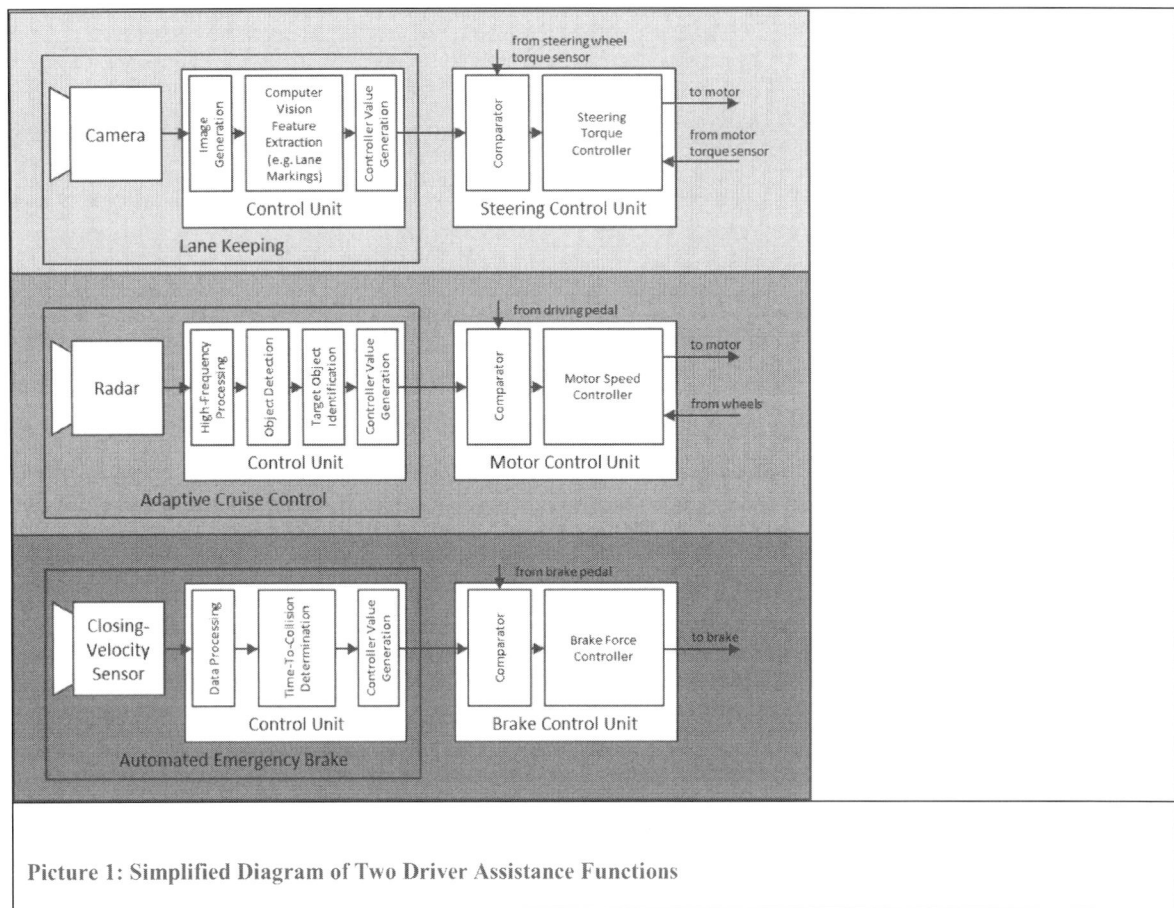

Picture 1: Simplified Diagram of Two Driver Assistance Functions

It is clear that this kind of close-to-sensor packaging helps use very little space. It also opens up the flexibility to market both functions to the end customer separately, giving a lot of flexibility in price and scaling. And last but not least, with this kind of separation it is easy to source the different systems from different vendors, which helps in negotiations to optimize price and performance. So in this evolution stage, system separation theoretically leads to finding the optimal solution both in technology, cost, and vehicle integration, giving the fundamental liberties of

- **package optimization** – picking the smallest and easiest-to-integrate package possible,
- **combinatorial optimization** – giving the end customer or one's own sales department the liberty to pick-and-choose from the DAS offering, and
- **vendor optimization** – picking and choosing from different vendors offering the same functionality, optimizing cost, functionality, established collaboration relationships, or all of the above.

The nice aspect of this situation is that it allows separate testing of each function, and very little integrating work between the functions is necessary. Essentially, speaking in the Landau notation for complexity that is established in computer science, the complexity for integrating **n** functions into one vehicle production process is **O(n)**.

It does, however, lead to a major hurdle in evolution of these systems. As long as the assumption is that both of these systems only assist, not replace, a human driver, and the driver activates lane keeping and distance keeping separately and can override their behavior, everything is fine. As soon as system evolution leads to the desire to link the lane keeping and distance keeping algorithms, because it is driving without human control, this kind of separation leads to great difficulties.

Obstacles For Traditional Driver Assistance Systems

Let us consider the case of a partially-automated driving system[2] which would, in the simplest way possible, combine the three driver assistance functions outlined in Picture 1 to form one complete system that drives in its own lane, performing emergency braking if necessary. Let us examine the following real use case:

The radar sensor picks up a vehicle driving in the vehicle's own lane, determines its velocity and finds it to be of similar speed of the ego vehicle[3]. As a result, the controller value generator sends a controller value equal to the current speed to the motor controller, which keeps the throttle at around the current position.

At the same time the closing-velocity sensor picks up an object which would require an emergency brake. This might be a pedestrian entering the path of the vehicle, which is not picked up by the radar, perhaps due to that sensor's restricted opening angle. There-

---

[2] sometimes called a "Level 2" system, referencing the levels of automation put forth by SAE and VDA. Partially-Automated, for the sake of this article, means that the driver is in control and responsible for the driving task, but that the vehicle handles lateral and longitudinal motion by itself.

[3] the "ego vehicle", for the sake of this article, is the vehicle controlled by the DAS function under review.

fore, the emergency braking function sends a command to the brake control unit to bring the vehicle to a stop.

As the brake control unit performs this task, the wheel speed is reduced. This gets reported to the motor control unit, which finds the vehicle decelerating while it should actually (to all of the motor control unit's knowledge) keep its current velocity. This causes the motor control unit to accelerate, possibly negating the effect of the brakes. Depending on whether the motor or the brakes are stronger, the outcome of this situation might result in a collision with the pedestrian, or with a standing vehicle whose motor is fighting full throttle against the locked brakes. Neither of these is a desirable situation.

Technical Integration Troubles

Quite naturally, this situation can be resolved, and routinely is resolved in today's systems. The easiest way is to handle conflicting input at the actuator level. In Picture 2, we have introduced motor and brake coordinator software modules in the motor control and brake control units, respectively. These will cause the motor to stop accelerating if the automated emergency braking system issues a brake command.

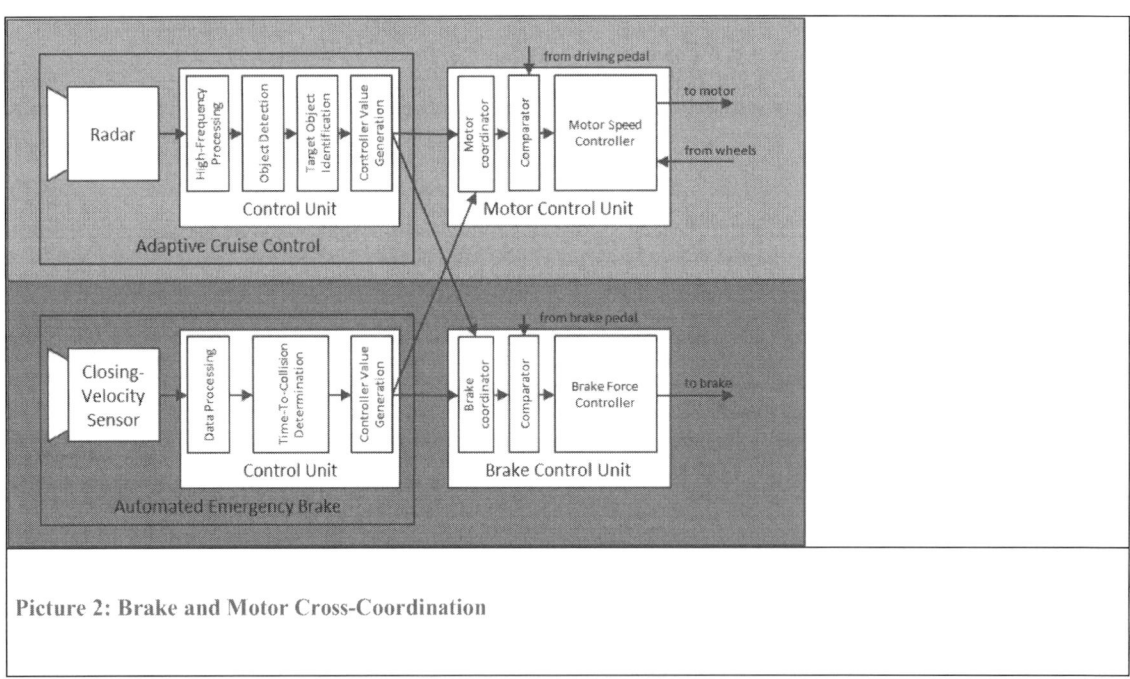

Picture 2: Brake and Motor Cross-Coordination

This is just a simple example. Taking it further, one might ask whether it is better to do the coordination at the function level (probably in addition, since at the actuator level possible function errors might be corrected as well), leading to more red arrows and more function blocks. The important factor is that the functions are no longer independent of each other now, and the interdependence needs to be very clearly specified to make sure that the compound behavior of cruise control and emergency brake is correctly executed. Also, if one wants to keep the benefit of system scalability to be able to sell the functions separately to the customer, correct behavior of the system now must be certified multiple times because the components influence each other when integrated into the same vehicle. For two such function interdependencies, we are now talking about three certification steps (for a vehicle with only function A, only function B, or function A+B working together).

If the liberty of **combinatorial optimization** is to be retained, the resulting complexity of certifying all of the possible combinations can be expressed as $O(2^k-1)$ for **k** systems. It is apparent that even if the interdependence could be resolved with very little work, and even if many of these combinations can in the end be pruned, an exponentiality to the power of **k** will very quickly run out of every bound.

Organizational Integration Troubles

Apart from the purely technical concerns, there is also a big organizational complexity to take into account. If one wants to keep the liberty of **vendor optimization**, this means that the technical work outlined above is, at least in responsibility, the work of the integrator instead of the function developer. All the possible interdependencies must be specified, traced, tracked, and tested by the integrator. In an ideal world, this would just be work that is to be done, regardless by whom. In the real world, however, every communication nexus means some loss of information, possibility for misunderstandings and misinterpretation of specification documents and annotations. This situation is more aggravated by the very real possibility that in a real project, vendor X delivering function A might also have a function B in its portfolio, however the integrator chose to buy function B from vendor Y. As a result, the integrator needs to coordinate the cooperation between the competitors, vendor X and vendor Y, who clearly are anything but natural co-operators. So if we could determine a "friction factor" **d** outlining the complexity that arises from getting several competing businesses to play nicely with each other and the integrator, we might write the complexity as $O((1+d)(2^k-1))$.

Of course the situation does not get better if we distribute the responsibility for the separate functions over multiple departments or groups in the integrator's own organizational structure, as is very often the case. We will not complicate the formula further, since this additional complexity can be expressed by simply making **d** a bit larger, but this enlargement should be kept in mind nonetheless.

Coping With Complexity

The current state of the art is quite beyond that outlined in the few simple functions above, but the integration complexity is very much at the stated level. There is one big trend in the automotive industry at the moment, and it is pointing towards **centralized control units**.

Can Centralized Control Units Help?

Essentially, all sensor inputs are collected in a centralized high-power computer, and all functionality is calculated on that computer. The result might look a bit like in Picture 3. For the sake of clarity, we are still using the very simple function examples from above. We kept the basic blocks of the Adaptive Cruise Control and Emergency Braking functions, but pulled them apart in sensor processing modules (which stay in the sensor control unit, for this example) and function modules (which move into the central control unit). Since we now have the sensor inputs all in one control unit, we added a sensor data fusion block, which improves on the data received from radar and CV sensor both. Whether this is needed or even useful is a matter that shall not be debated here.

One main argument for centralized control units that has been quoted often is that they offer superior computational power. In recent realizations of this concept, this has often been true, but why it would not be possible to bring this computational power into another control unit which would already be existing in the vehicle has not been thoroughly explored. Since we are focusing on essential liberties and complexity here, computational power shall not be considered in the context of this paper.

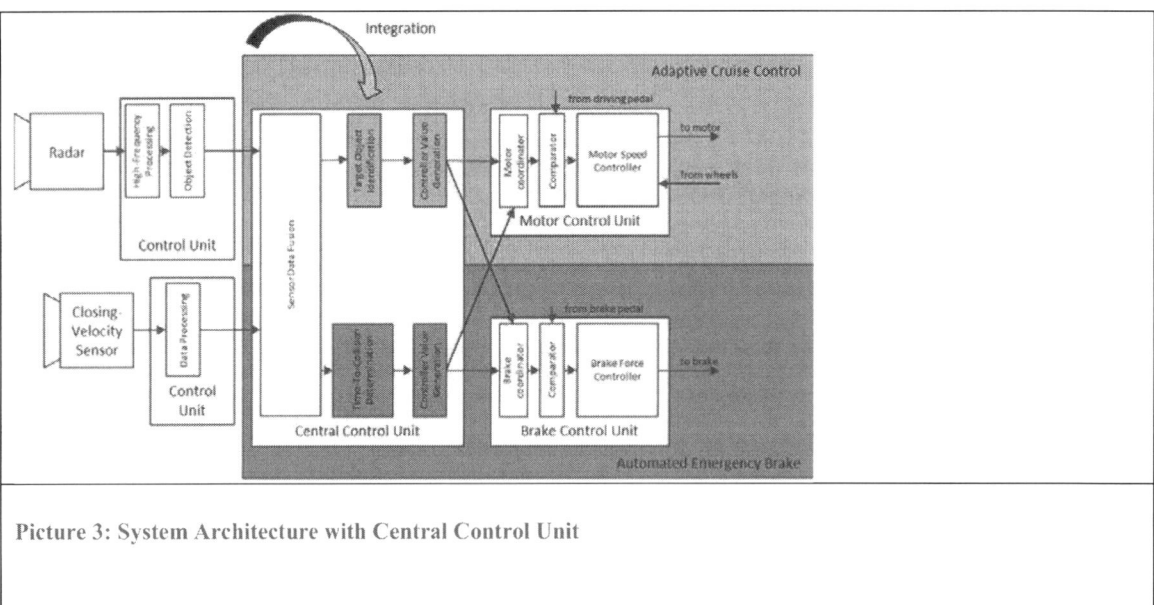

Picture 3: System Architecture with Central Control Unit

Does this central control unit improve on the development complexity and our fundamental liberties? Let's see.

One important thing that has been achieved is the liberation of the functional software modules from its accompanying hardware. It is now possible to buy the function modules running on the central control unit from a different vendor than the one who sells the radar and CV sensor CUs. The liberty of vendor optimization has been improved upon. That of package optimization has not, however. After all, we now have one control unit more in the vehicle than before. This might be an investment that evens out if more functions are integrated than just one, so it may be not a significant drawback. What has happened to combinatorial optimization? Can we pick and choose functions as we please?

Yes, we can, naturally. However, we must pay for that central control unit every time, as it must be developed in such a way to fit all possible functions developed for a certain vehicle model line, in terms of RAM, ROM, processing power, and bus connectors. We might assume there would be some scalability involved, e.g. leaving some chips unpopulated for scaled-down functionality, but the PCB, enclosure, room for connectors (both in the ECU's casing and in the space it will fit into in the vehicle), are hard to scale. So we might argue that we actually have lost on combinatorial optimization by including this central control unit.

How about complexity? At the very least, someone has to integrate all the software modules into the new ECU. This is (to formulate a lower bound) at least not less work than integration into the existing control units, so we can say for certain that complexity has not improved. An interesting side question is who might be the party doing the integration. It might be the producer of the central control unit, who may or may not be also the vendor of one of the sensors, or it might be a third party (including the OEM themselves). In the former case, we might see preferred integration support for the vendor's own code and possibly worse integration support (true or alleged) for the other parties' code. In the latter case, we would see increasing complexity by involvement of another party, due to the increasing number of edges in the communication graph. In any case, complexity is certainly not diminishing.

Concluding, use of a centralized control unit trades an increase in vendor independence for degradation in package and combinatorial optimization, and (in the best case) no change in complexity. One plus and two or three minuses does not seem to make this a very attractive option in terms of strategic, architectural freedom.

The Role of Software

So far, we have looked at the software modules we are working with as rather passive assets, that are implemented and integrated in certain control units. However, examining the pictures, it becomes clear that much of the complexity is actually handled not by hardware but by software. Managing the underlying hardware network is a compelling task in itself, but it is aided by the fact that at least the communication hardware and low-level protocols, such as CAN, FlexRay, or Ethernet, are well-defined and handled by standardized transceiver modules. Therefore it is the software components we should focus on.

Handling Complexity

Coping with highly complex architectures has been a domain of software development from the very beginning, and a number of mechanisms have been developed to help with it. We will examine some of them and see what contribution they can have to reduce system complexity in an automotive context.

Abstraction

One of the core principles in software development is abstraction. If a software component needs to co-ordinate too many partners, it becomes a good idea to hide this behind a newly-introduced component which is solely responsible for this co-ordination. A good example from standard software programs is handling mass storage. A program could interface directly with the hard disk and address individual sectors, but then it would have to determine which sector is in use by another program, or possibly defective. To handle this, the idea of a file system is introduced. The file system presents a programming interface to every program which abstracts away all those tasks, and also can handle storing files on the network or (temporarily) in RAM.

Let us revisit the two-function example from above.

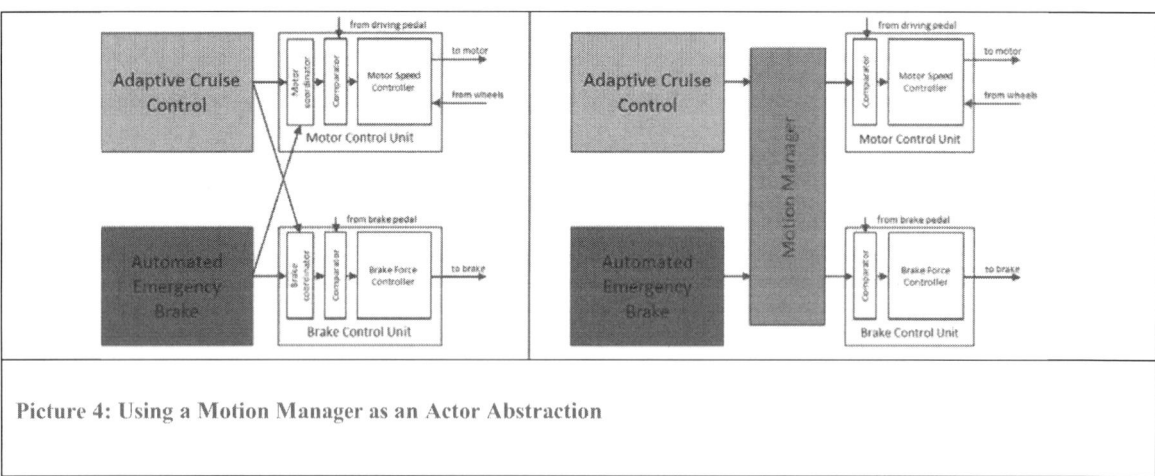

Picture 4: Using a Motion Manager as an Actor Abstraction

We have introduced a new component, called (for the sake of discussion) **Motion Manager**, which presents an abstraction to the functions needing to access the motor and brake control units, respectively. This component accepts acceleration and deceleration requests from the functions, decides with its own internal logic which request to honor and how to honor it (e.g. by stalling the motor or by braking), and then co-ordinates communication to the actor units.

How does this change complexity? One might argue that we have now introduced another component into the system. However, both the functions and the actors now have one communication channel instead of two, and close examination shows that the brake and motor coordinator blocks have been eliminated, or rather, moved into the Motion Manager as part of that component's internal logic. So while the new component is certainly more involved than each of the two coordinator components in the left image, one might argue that in reality, two components have been eliminated for the price of adding one more. The Motion Manager itself has one communication channel to each function and one to each actor, so its complexity scales linearly with each additional component.

Revisiting the complexity examination presented above, we have gone from $O(2^k-1)$ to $O(k)$, moving from exponential complexity to linear complexity in this part of the system. For the sake of the example, the Motion Manager in Picture 4 has been kept very simple. In a real vehicle, it can do much more with only linear effort: Decide whether to accelerate with the electrical or combustion engine of a hybrid vehicle, or both; decide whether a brake request must be handled by friction braking or whether recuperative braking, stalling, or simply coasting up an incline is sufficient, and (when enhanced to co-ordinate lateral motion as well) whether to steer with front and/or back axle or by one-sided braking. There is quite some potential for redundancy in the control of a state-of-the-art vehicle, and the Motion Manager can help exploit it, for example for functional safety purposes, or even help clarify if some of it can be eliminated.

Other possibilities for complexity reduction can be sensor data fusion components, HMI co-ordinators, supervisor components such as required by functional safety, and many more.

Polymorphism

One core concept of object-oriented system design is the idea that software components that do similar things, even if they differ very much in their internal structure, should present the same interface to the outside. This is called Polymorphism, and it is a principle that can be exploited in vehicle architecture as well.

So far we have looked at the components making up a vehicle system, but the communication edges bear quite some potential for effort reduction as well. Looking at the hardware components in Picture 2, both of the functions will send deceleration requests to the brake. In today's vehicle architectures, it is quite common to use different interfaces on both these communication edges. This has technical merit: One function might require finer granularity in deceleration, while another one might have higher functional safety demands. Using the same interface for both channels might waste some communication bits, which on a crowded CAN network are an expensive resource. But as the industry moves towards more complex function architectures and higher-bandwidth bus systems such as FlexRay or Ethernet, the scales tip in the direction of the notion of simpler, standardized interfaces.

The great advantage in this is that the developer of a component interfacing with several partners issuing similar requests does not need to write new code to honor each request. There might be differences between the requests, which might be encoded in flags or priority indicators, but the fundamental way to handle each request stays the same. It is now possible to use the same (slightly more complicated) program code to interface with many partner components, bringing the complexity in this part of the system down from **O(k)** to **O(1).**

Standardization

Another aspect of putting some more work into the interface specification to reduce code complexity is that of standardization. In his seminal work on object-oriented programming [2], Brad Cox likens software components (which he calls *software ICs*) to the standardized TTL logic ICs of the 74xx family, which were introduced by Texas Instruments in the 1970s. Nowadays, a 74HC11 chip (a triple AND gate) is available from many vendors, and its logic is realized internally in quite different ways, but its interface and the contract of its functionality is always the same.

Interestingly, if the work required for polymorphism as described above is done, the most important step for standardization is already finished. Of course, we are still discussing one vendor's system, so this is a standard which can be exploited over several components in the same system, and maybe over time.

But what were to happen if the interfaces for a certain type of component (say, a driver assistance function handling longitudinal control) were not just set for one vendor or OEM but instead published and adopted over several OEMs? This would greatly increase competition, since now it would be possible to replace one vendor's function by another one in the next vehicle generation. This in turn would help lower development costs, and reduce the risk of vendor lock-in. Both of these points are advantages for the OEM and not so much for the component vendor. Another advantage would be that a standardized component provided by one vendor could be used across several OEM's systems, which in turn greatly reduces development cost for the vendor. If functional safety and certification aspects are pulled into view, which may result in the largest cost factor in bringing highly-automated driving systems to market, this opens the possibility of a much higher return on investment for both the component vendor and the OEM.

Of course, this kind of standardization is commonly associated with long and involved committee work. Interestingly, to keep with the example of the 74xx family of TTL logic ICs, there exists no formal standard for these components. Simply put, these were first produced by Texas Instruments, and documented well enough, and their design was never patented. Other companies adopted the interface specification (i.e. pin-out) and form factor and competed with Texas Instruments on price and quality issues such as EMV safety, leakage current and voltage requirements, and a quasi-standard emerged that exists to this day.

Examples for this kind, and more formal kinds, of standard exist in the automotive world: AUTOSAR [3], ADASIS [4], and the consortial process that is beginning at the time of publication of this article for cloud-based mapping architectures, SENSORIS [5].

Robot Architectures

Finally, let us examine the field of autonomous robotics, where the scientific community has exploited the above topics to the maximum. A highly-automated vehicle is nothing else, in essence than an autonomous robot that must perform sequences of complex behaviors to be able to successfully negotiate road traffic. A very simple sequence of actions might be

- follow a leading vehicle and simultaneously keep in the lane

- if the vehicle in front becomes too slow, and there is space in the adjacent lane, perform a lane change

- once the vehicle in front has been passed, perform a lane change to the previous lane again

- if the current lane is blocked by an obstacle, and a lane change is not possible in time, perform an emergency braking manoeuver

In terms of technological requirements, this means that the vehicle, at the least, needs to have a clear idea of the structure of the environment (obstacles, lane markings, moving vehicles), and it needs to be able to control its speed and steering in a comfortable and possibly, in an emergency, un-comfortable way.

It is clear that all these actions must be performed on a common view of the environment, and must be able to control the vehicle actors in a concise and safe way. As has been shown above, the abstractions of Motion Manager (for vehicle control) and Sensor Data Fusion (to obtain an environment model) can be used to reduce the complexity induced by the multitude of sensors and control components in the vehicle. Picture 5 shows this in principle.

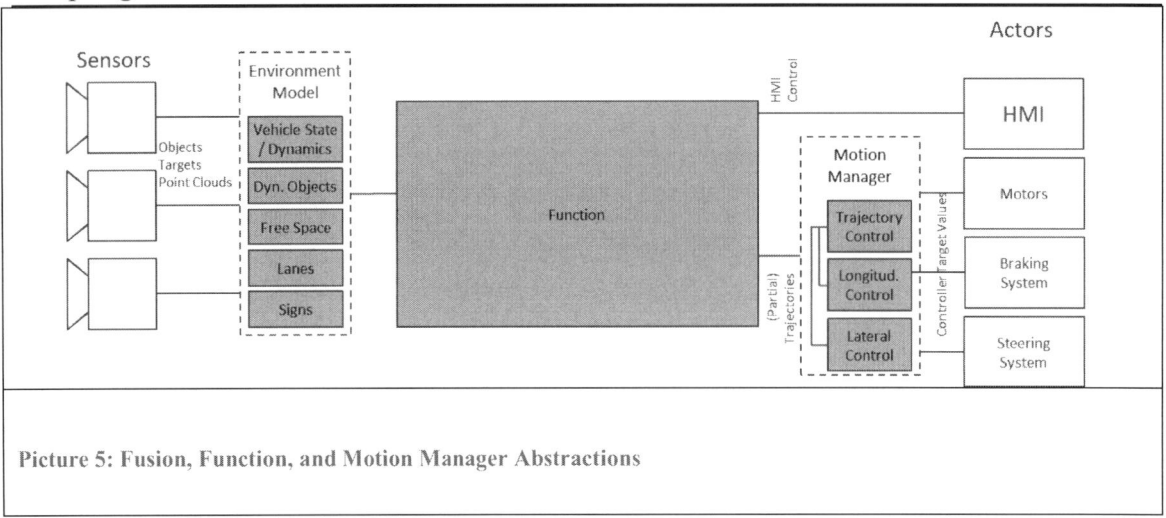

Picture 5: Fusion, Function, and Motion Manager Abstractions

However, this still leaves a very big function block in the middle which in itself is exceedingly complex to develop. Can it be cut into smaller pieces that reduce this complexity?

Subsumption Architectures

This question has been addressed in autonomous robot research since the 1980s, when Rodney Brooks and a team of students developed the subsumption robot architecture at MIT. [6] The state-of-the-art robot systems at that time worked by analyzing commands, formulating and then executing a plan, but had problems changing that plan if they encountered obstacles that proved the original plan to be inoperable.

Frustrated by the difficulty to develop more complex, fault-tolerant behavior in this kind of system, Brooks developed an architecture that broke down robot actions into several simplified behaviors, and used simple rules to coordinate which behavior could operate at any given time. Picture 6 shows a simple example of a robot that can explore the world according to a plan, simply wander around if a plan cannot be followed, and avoid obstacles at the same time.

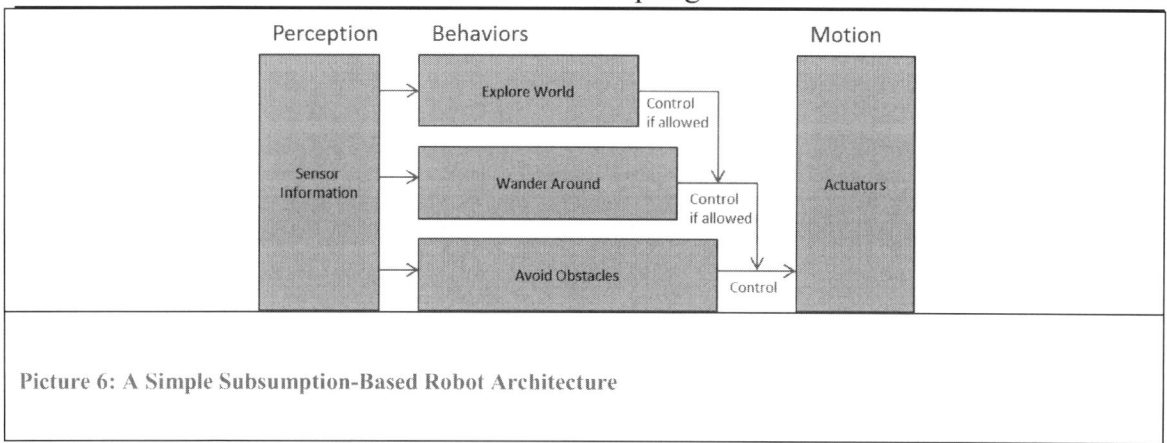

Picture 6: A Simple Subsumption-Based Robot Architecture

The idea of the subsumption architecture is that the robot has low-level behaviors that are very simple (such as "avoid obstacles") in Picture 6, and more complex ones that "subsume" the lower-level behaviors, aggregating their expertise and adding more. This is an alluring concept for intelligent robots; however, it is clear that if the lower-level behavior changes in a significant way, the higher layers must be adapted to the new state of affairs as well. Also, in this simple architecture, all behaviors are active all the time, which may cause conflict that then must be resolved (and also is unnecessarily resource-intensive). One bane of subsumption-based architectures is often that the behavioral network has a tendency of becoming unmanageable as behavioral complexity increases.

More General Behavior-Based Architectures

Following the relative success of subsumption architectures but needing more complex control mechanisms, a number of different approaches emerged. One is the idea of activating behaviors that govern different parts of the robot at the same time, then fusing their actuator outputs. In [7], this is used to control a walking machine. One simple example in this setup might be the combination of one behavior that raises a leg, another one that swings it forward, and a fusion block that combines both these behaviors. Through an activation output, the behaviors can exert some control over the outcome of the fusion.

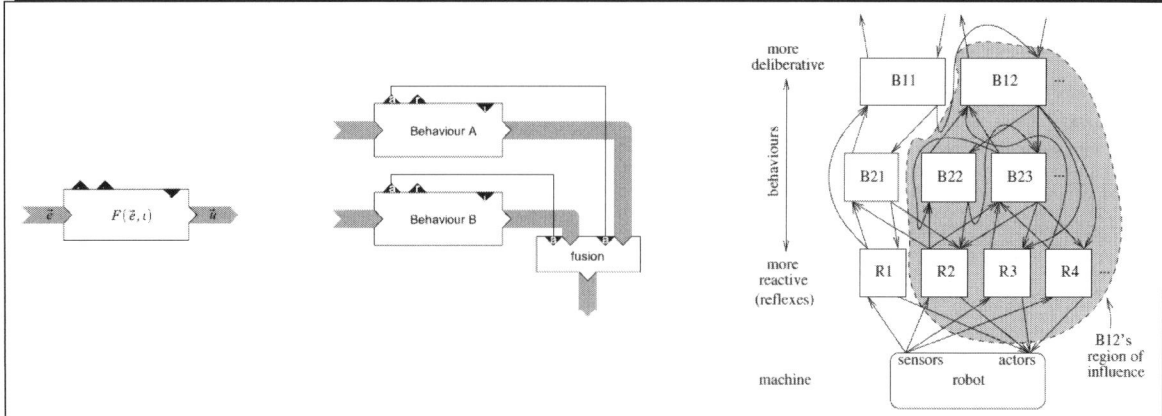

Picture 7: Elementary Behaviors, Activation, Fusion, And Networking Behaviors *(Sourc: Albiez et al., A Behaviour Network Concept for Controlling Walking Machines, In: Adaptive Motion of Animals and Machines. Springer Tokyo, 2006. S. 237-246.)*

Here, as in the basic subsumption architecture, there exists a hierarchy of more simple and more complex behaviors, although the idea of subsumption (as in integration of simpler elements) has been dropped in favor of one of utilization. Very interesting is the fact that the behaviors have *standardized interfaces:* an activation / inhibition input, an activity level output and a rating output (the latter is essentially a way of saying "I am certain (or uncertain) that what I am currently doing is the right thing").

With these simple interfaces, it is possible to build, manage, and monitor very complex system behavior.

Adapting Robot Architectures For Vehicles

The very flexible but detailed architectural examples outlined above were developed as abstract architectural models without a specific robotic application in mind. That is to say, they can be used to program automatic lawn mowers just as well as autonomous robots repairing a space station. Of course, this huge flexibility is not what comes to mind when talking about autonomous vehicles. It is neither desirable nor required. After all, the task to be performed by an autonomous vehicle is rather specialized and clearly defined, we can assume a lot (but not everything) about the environment the system is going to operate in, its sensors and actuators.

Still, the behavior it needs to exhibit can get quite complex, as outlined above; the basic principle of breaking down the complexity into smaller, manageable parts can therefore applied to some profit. Consider Picture 8, which is an expanded version of Picture 5.

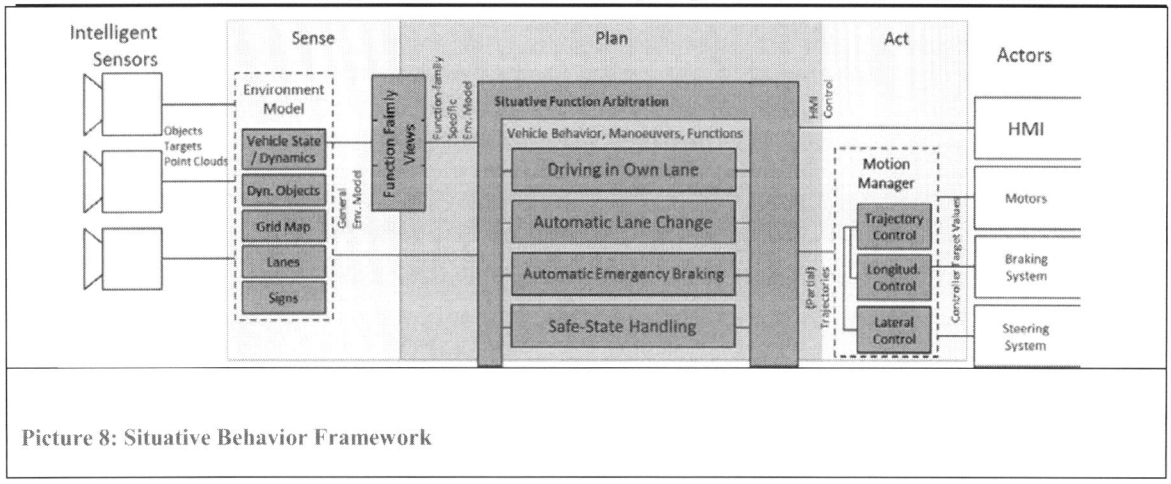

Picture 8: Situative Behavior Framework

Here, we have introduced a number of simple behaviors that reflect the highway-driving scenario outlined on page 11: Driving while keeping in the car's own lane, performing automatic lane changes, emergency braking if needed. We have added a behavior called "safe-state handling", which might be activated in case of system failure and could direct the car to a complete stop on the road shoulder. It is interesting to note that these behaviors do not necessarily need to be re-developed. For example, automated cruise control, lane keeping, and automated emergency braking are well-established functions in today's cars. They will function within a behavioral architecture just as well; only their interfaces change.

Since automated driving is wrought with considerable safety requirements, it is a good idea not to give these behaviors access to the motion manager component at the same time but rather embedding them in an arbitration block, which governs, monitors and secures interaction of the behaviors. This arbitration block can fulfill another role: It essentially decides when which function is allowed to operate, and it does this according to a set of programmed rules. However, these rules might change depending on the situation. If the vehicle is currently being steered by the driver, things like automated lane change or safe-state handling might be deactivated, and the threshold for automated emergency braking might be raised (since the car does not know if the driver might swerve to avoid an obstacle). If the driver yields control to the automation system, all blocks become activatable, and the automated emergency braking's threshold gets lowered (since the car plans its full path, so if it looks like a collision is about to occur, the software must have done something wrong before). These different behavior sets can be implemented in rule sets inside the arbitration framework.

To revisit our complexity argument, we have now reduced the number of communication partners for each behavior to four: The environment model, the motion manager, the HMI, and the arbitration framework. Due to the standardized interfaces used in the architecture, this is it; regardless of the number of behaviors, the complexity stays the same. We have therefore limited the complexity for the function developer to **O(1)**.

Software Distribution and AUTOSAR

It should be noted that all the discussion about software abstraction mechanisms has been centered purely on software architectures. Control units have not been the center of attention. But software needs to run on hardware. How does partitioning work?

It has taken far more time for the AUTOSAR base architecture to really take hold and become pervasive in the automotive industry than originally expected. But now that it has, it can reasonably be expected that most of the control units in a given car project are running some form of AUTOSAR. But AUTOSAR started life with the promise of decoupling software from hardware. Industry-wide, this has not happened so far; ECUs are running the AUTOSAR base software, but proprietary, compartmentalized application software on top of it. There are certainly many reasons for this, but one might be that the application software was never described in a sufficiently generic way that it was actually reasonable to move a component from one ECU to the next. For driver assistance, robot architectures have the potential to change this.

Not only can they specify a clear, functional architecture, but also the interfaces between all the software components. The AUTOSAR specification language, ARXML, is ideal to describe these interfaces. This is great news for software modules in standardized system architectures, since it allows clear description, extendibility, and portability – from one project to the next within a manufacturer's production chain, but also between suppliers and manufacturers. AUTOSAR is therefore the ideal base-level operating system for a robot architecture. It allows distribution of specified software components to real-world control unit networks, essentially *mapping the software architecture to hardware*. This mapping step is not trivial, since it is essentially an optimization task keeping in mind constraints such as control unit capabilities (processor speed, ROM, and RAM), networking bandwidth, and functional safety matches between a control unit and the software running on it. But it can be done.

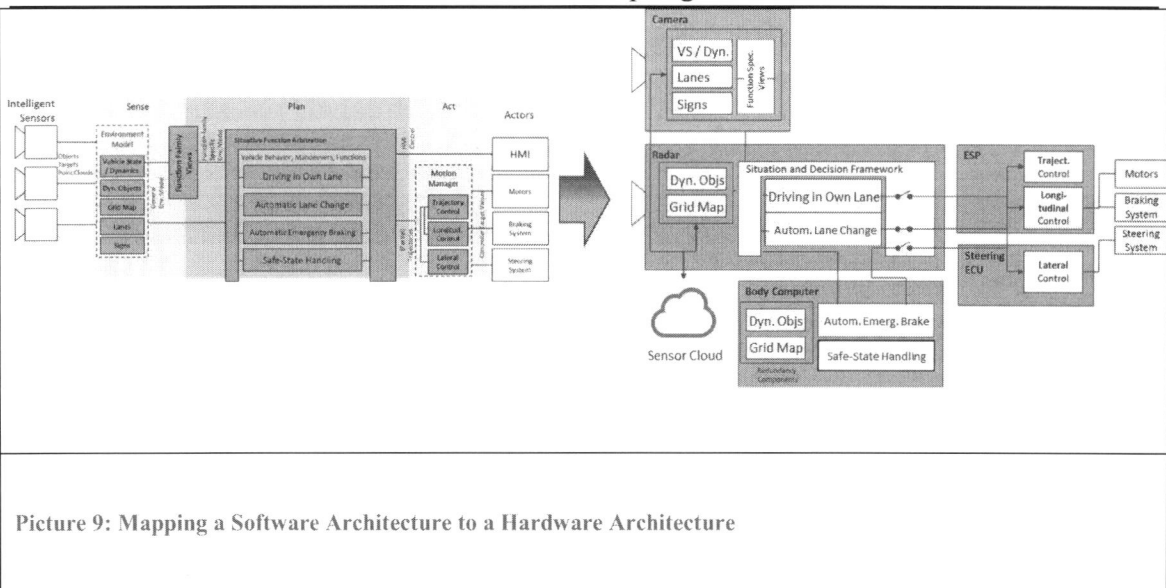

Picture 9: Mapping a Software Architecture to a Hardware Architecture

Finally, to come back to our fundamental liberties of package optimization, combinatorial optimization, and vendor optimization, it can be said that:

- **For package optimization**, an architecture as the described one is optimal as it can use the hardware that exists in the vehicle. The mapping step can make sure that all the available processing power across the vehicle can be used (honoring technical constraints, of course);
- **For combinatorial optimization**, it is optimal because it is fundamentally scalable – functions or behaviors can be left out or put in at will, and can be distributed across control units so that leaving out one function and leaving out one control unit correspond;
- **For vendor optimization**, it is optimal because components are standardized – picking and choosing from different vendors offering the same functionality, optimizing cost, functionality, established collaboration relationships, or all of the above, is made easy.

Conclusion

In this paper, we have examined some of the biggest hurdles for development of such highly complex vehicle systems as automated driving. The main hurdle is just that complexity, as well as the fact that so far, few mechanisms exist or are in use in the automotive industry to handle it. To cite one example, the current trend towards centralized control units does certainly help in a number of aspects (computational power, networking bandwidth between components), but it does not help reduce complexity. We have also discussed some mechanisms from software engineering that do help if rolled out on a system level, and introduced behavioral robot architectures as a wide-reaching example of these mechanisms.

We believe it is time to tackle these complexity issues on an industry-wide scale, moving towards standardized systems architectures and software modules, to make development of highly automated vehicles easier, and in the end, cost-effective.

Bibliography

[1] K. Bengler, K. Dietmayer, B. Färber, M. Maurer, C. Stiller and H. Winner, "Three Decades of Driver Assistance Systems," *IEEE Intelligent Transportation Systems Magazine,* vol. 6, pp. 6-22, 2014.

[2] B. Cox, Object-Oriented Programming: An Evolutionary Approach, Addison Wesley, 1991.

[3] AUTOSAR consortium, "http://www.autosar.org," [Online].

[4] ADASIS forum, "http://www.adasis.org," [Online].

[5] D. Rabel, "SENSORIS – Sensor Ingestion Interface Specification – Closing the sensor data feedback loop," in *Tech.AD*, Berlin, 2016.

[6] R. Brooks, "A Robust Layered Control System For a Mobile Robot," *IEEE Journal of Robotics and Automation,* vol. 2, no. 1, p. 14–23, 1986.

[7] J. Albiez, T. Luksch, R. Dillmann and K. Berns, "A Behaviour Network Concept for Controlling Walking Machines," in *Adaptive Motion of Animals and Machines*, Tokyo, Springer, 2006, p. 237–244.

Fault-tolerant components for automatic driving automobiles – some basic structures and examples

Prof. Dr.-Ing. Dr. h.c. Rolf Isermann

Institute of Automatic Control and Mechatronics
Technische Universität Darmstadt

© Springer Fachmedien Wiesbaden GmbH, ein Teil von Springer Nature 2018
R. Isermann (Hrsg.), *Fahrerassistenzsysteme 2016*, Proceedings,
https://doi.org/10.1007/978-3-658-21444-9_14

1 Introduction

Partially and highly automated driving automobiles are characterized by automatic controlled longitudinal and lateral movement and at least a reduced, permanent acting driver. This requires an increased supervision of all active systems and a fault-tolerant design of the safety-relevant components of the chassis and the powertrain.

As for all *safety-related systems*, all aspects of reliability, availability, maintainability and safety (RAMS) have to be considered. To meet safety requirements, special procedures were developed in different technical disciplinces like railway, aircraft, space, military, nuclear and, later, automotive systems [1]. These procedures are covered by the terms *system integrity* or *system dependability*. Safety and reliability are generally achieved by a combination of: fault avoidance, fault removal, fault tolerance, fault detection, fault diagnosis, automatic supervision and protection.

To investigate the effect of faults on the reliability and safety *during the design* and type certification, a range of analysis methods were developed, mainly reliability analysis, event tree analysis (ETA), fault tree analysis (FTA), failure mode and effect analysis (FMEA), hazard analysis (HA) and risk classification, see e.g. [2–4].

The *reliability* can be improved by oversizing, maintenance, protection for mechanical, hydraulic, electrical and some electronic components and by redundancy especially for electrical components, electronic hardware and software.

Because not all faults and failures can be avoided totally after careful design, manufacturing and testing, high-integrity systems must have the ability of *fault tolerance*. This means that faults are compensated in such a way that they do not lead to system failures.

2 Fault tolerance for components

Fault-tolerance methods generally use *redundancy*. This means that in addition to the considered module one or more modules are connected, usually in parallel. These redundant modules are either *identical* or *diverse*. Such redundant schemes can be designed for hardware, software, information processing, mechanical and electrical components, like sensors, actuators, microcomputers, buses, power supplies, etc.

There exist mainly two basic approaches for fault tolerance, static redundancy and dynamic redundancy. The corresponding configurations are first considered for *electronic hardware* and then for other components, [4], [5].

Figure 1a) shows a scheme for *static redundancy*. It uses three or more parallel modules which have the same input signal and are all active. Their outputs are connected to a voter who compares these signals and decides by majority which signal value is the correct one.

2

If a triple modular redundant system is applied and the fault in one of the modules generates a wrong output, this faulty module is masked (i.e. not taken into account) by the 2-out-of-3 voting. Hence, a single faulty module is tolerated without any effort for specific fault detection. n redundant modules can tolerate $(n-1) / 2$ faults (n odd).

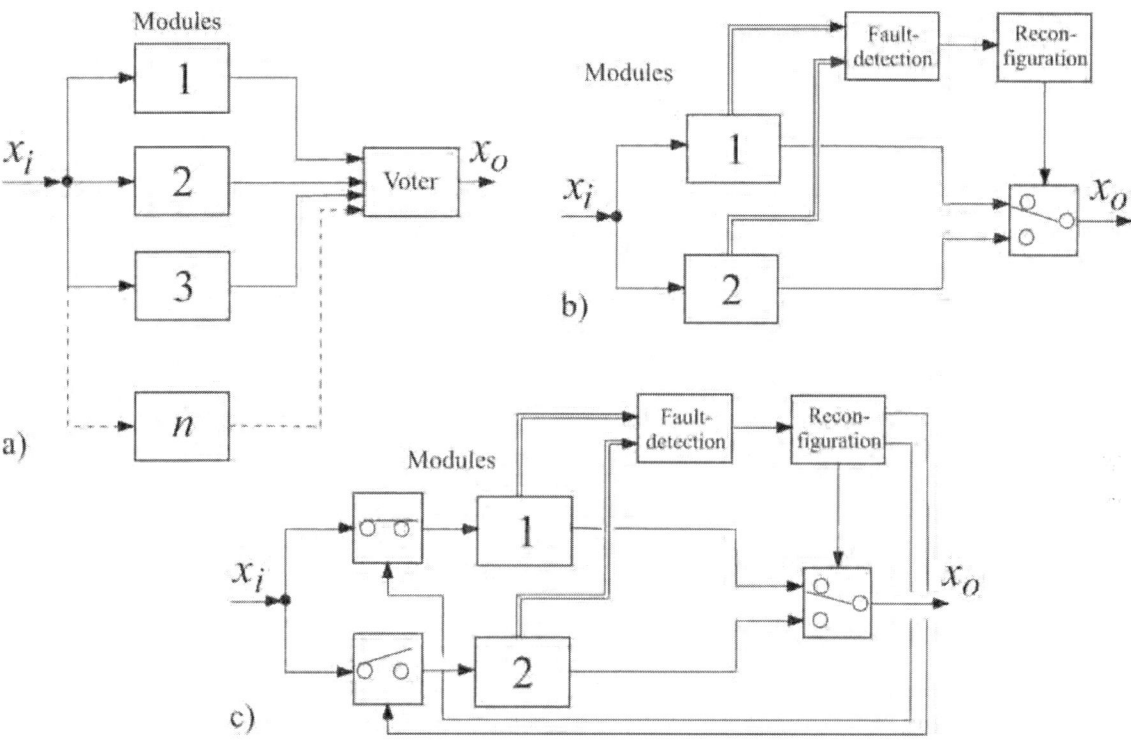

Figure 1. Fault-tolerant schemes for electronic hardware: a) Static redundancy: multiple redundant modules with majority voting and fault masking, m out of n system (all modules are active); b) Dynamic redundancy: Standby module which is continuously active, "hot standby"; c) Dynamic redundancy: Standby module that is inactive, "cold standby".

Dynamic redundancy needs less modules on cost of more information processing. A minimal configuration consists of two modules, Figure 1b) and c). One module is usually in operation and if it fails the standby or backup unit takes over. This requires a fault detection to observe if the operation modules become faulty. Simple fault-detection methods use the output signal only for, e.g. consistency checking (range of the signal), comparison with redundant modules or use of information redundancy in computers like parity checking or watchdog timers. After fault detection it is the task of the reconfiguration to switch to the standby module and to remove the faulty one.

In the arrangement of Figure 1b) the standby module is continuously operating, called "*hot standby*". Then, the transfer time is small on cost of operational aging (wear out) of the standby module. Dynamic redundancy, where the standby system is out of function and does not wear, is shown in Figure 1c), called "*cold standby*". This arrangement

3

needs two more switches at the input and more transfer time due to a start-up procedure. For both schemes the performance of the *fault detection* is essential.

Similar redundant schemes as for electronic hardware exist for *software fault tolerance*. Here, tolerance against mistakes in coding or errors of calculations is meant. The simplest form of a *static redundancy* is repeated running ($n \geq 3$) of the same software and majority voting for the result. However, this only helps for some transient faults. As software faults in general are systematic and not random, a duplication of the same software does not help. Therefore, the redundancy must include *diversity of software*, like other programming teams, other languages, or other compilers. With $n \geq 3$ diverse programs a multiple redundant system can be established followed by majority voting as Figure 1a). However, if only one processor is used, calculation time is increased, and using n processors may be too costly.

Dynamic redundancy by using standby software with diverse programs can be realized by using recovering blocks. This means that in addition to the main software module other diverse software modules exist, [3], [6].

Fault tolerance can also be designed for purely *mechanical* and *electrical systems*. Static redundancy is very often used in all kind of homogeneous and inhomogeneous materials (e.g. metals and fibers) and special mechanical constructions like lattice-structures, spoke-wheels, dual tyres or in electrical components with multiple wiring, multiple coil windings, multiple brushes for DC motors, multiple contacts for potentiometers. This quite natural built-in fault tolerance is generally characterized by a parallel configuration. However, the inputs and outputs are not signals but, e.g. forces, electrical currents or energy flows and a voter does not exist. All elements operate in parallel and if one element fails (e.g. by breakage) the others take over a higher force or current, following the physical laws of compatibility or continuity. Hence, this is a kind of "stressful degradation". Mechanical and electrical systems with dynamic redundancy as depicted in Figure 1b), c) can also be built. Both cold or hot standby may be applied.

Fault tolerance with dynamic redundancy and cold standby is especially attractive for *mechatronic systems* where more measured signals and embedded computers are already available and therefore fault detection can be improved considerably by applying process-model-based approaches. Table 1 summarizes the appropriate fault-tolerance methods for the case of electronic hardware.

Mainly because of costs, space and weight, a suitable compromise between the degree of fault tolerance and the number of redundant components has to be found for *automotive systems*. In contrast to fly-by-wire systems, only one single failure usually must be tolerated (presently) for hazardeous cases, [7], mainly because a safe state can be reached easier and faster. This means that not for all components of automobiles very stringent fault-tolerance requirements are needed. Following steps of degradation are distinguished:

- *Fail-operational* (FO): One failure is tolerated, i.e. the component stays operational after one failure. This is required if no safe state exists immediately after the component fails,
- *Fail-silent* (FSIL): After one (or several) failure(s) the component behaves quiet externally, i.e. stays passive by switching off and therefore does not influence other components in a wrong way,
- *Fail-safe* (FS): After one (or several) failure(s) the component possesses directly a safe state (passive fail-safe, without external power) or is brought to a safe state by a special action (active fail-safe, with external power).

Table 1. Fail behavior of electronic hardware for different redundant structures. FO: fail-operational; F: fail; (FS: fail-safe not considered)

| Structures | Number of elements | Static redundancy | | Dynamic redundancy | | |
		Tolerated faults	Fail behavior	Tolerated failures	Fault behavior	Discrepency detection
Duplex	2	0	F	0	F	2 comparators
				1	FO-F	fault detection
Triplex	3	1	FO-F	2	FO-FO-F	fault detection
Quadruplex	4	1	FO-F	3	FO-FO-FO-F	fault detection
Duo-Duplex	4	1	FO-F	-	-	-

For vehicles it is proposed to subdivide FO in "long time" and "short time". Considering these degradation steps for various components one has first to check if a safe state exists. For automobiles (usually) a safe state is stand still (or low speed) at a non-hazardous place. For components of automobiles a fail-safe status is (usually) a mechanical backup (i.e. a mechanical or hydraulic linkage) for direct manipulation by the driver. Passive fail-safe is then reached, e.g. after failure of electronics if independent of the electronics the vehicle comes to a stop, e.g. by a closing spring in the throttle or by actions of the driver via mechanical backup. However, if no mechanical backup exists as for drive-by-wire steering or braking after failure of electronics, only an action by other electronics (switch to a still operating module) can bring the vehicle (in motion) to a safe-state. This requires redundancy and the availability of electric power.

Generally, a graceful degradation is envisaged, where less critical functions are dropped to maintain the more critical functions available, using priorities, [2]. Table 1 shows degradation steps to fail-operational for different redundant structures of electronic hardware. As the fail-safe status depends on the controlled system and the kind of components, it is not considered here.

For *flight-control* computers usually at least a triplex structure with dynamic redundancy (hot standby) is used, which leads to FO-FO-FS, such that two failures are tolerated and a third one allows the pilot to operate manually [8], [36]. If the fault tolerance has to cover only one fault to stay fail-operational (FO-F) a triplex system with static redundancy or a duplex system with dynamic redundancy is appropriate. If fail-safe can be reached after one failure (FS), a duplex system with two comparators is sufficient. However, if one fault has to be tolerated to continue fail-operational and after a next fault it is possible to switch to a fail-safe (FO-FS), either a triplex system with static redundancy or a duo-duplex system may be used. The duo-duplex system has the advantages of simpler failure detection and modularity.

3 Fault tolerance for control systems

For automatically controlled systems the appearance of faults and failures in the actuators, the process and the sensors will usually effect the operating behavior. With *feedforward control* generally all small or large faults influence the output variables and therefore more or less the operation. If the system operates with *feedback control*, small additive or multiplicative faults in the actuator or process are in general covered by the control actions because of usual robustness properties. This property is therefore a *passive control-loop fault-tolerance*. However, permanent additive and gain sensor faults will immediately lead to deviations from the reference values. For large changes of the behavior in actuators, process and sensors the dynamic control behavior becomes either too sluggish or too less damped or even unstable. Then, either a very robust control system or an *active fault-tolerant control system* is required to save the operation. In the last case it consists of fault-detection methods and a reconfiguration mechanisms which modifies the controller. Dependent on the kind of faults, the reconfiguration may change the structure and/or parameters of the controller. This can also include the switch to other manipulated variables or actuators or sensors, if available. However, fault detection in closed loops needs careful consideration [4].

6

Examples are fault-tolerant flight control system with reconfiguration to other control surfaces after failure of actuators of ailerons, elevators and rudders, see e.g. [8–11]. Fault-tolerant control for lateral vehicle control is treated in [12].

4 Fault management

In the frame of automated systems a fault tolerance can be designed for several components as actuators, sensors, communication channels, microcomputers, and control functions. Figure 2 illustrates a general automatic fault-management system with two manipulated and two controlled variables. The second actuator, for example, replaces the first actuator in the case of a fault if a similar manipulation effect can be reached. Hence, the components can be identical or diverse or redundant in itself. The fault-detection module operates in closed loop without or with test signals. The fault management then imitates a reconfiguration to the redundant component. This can occur initiates by hard or soft switch-over, change of operation status or switch to feedforward control. For more details see [13].

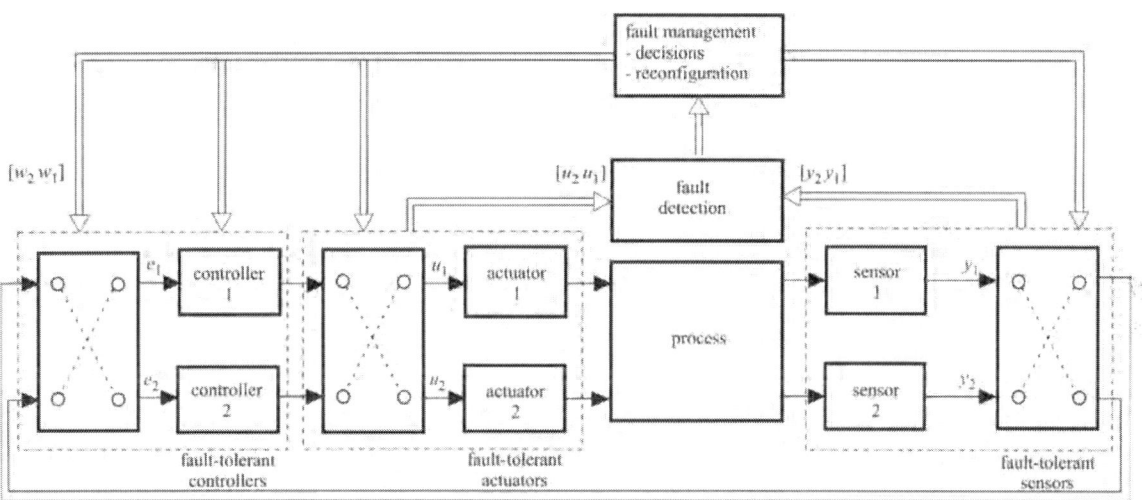

Figure 2 Fault-tolerant control system with automatic fault management, shown for two actuators, two sensors and two controllers.

5   Fault detection methods

Fault-detection methods based on measured signals can be classified in

- *limit value checking* (thresholds) and *plausibility checks* (ranges) of single signals,
- *signal-model-based methods* for single periodic or stochastic signals,
- *process model-based methods* for two and more related signals.

Figure 3 shows a scheme for these methods. For a description of the various methods it is referred to the literature, e.g. [4], [13–15].

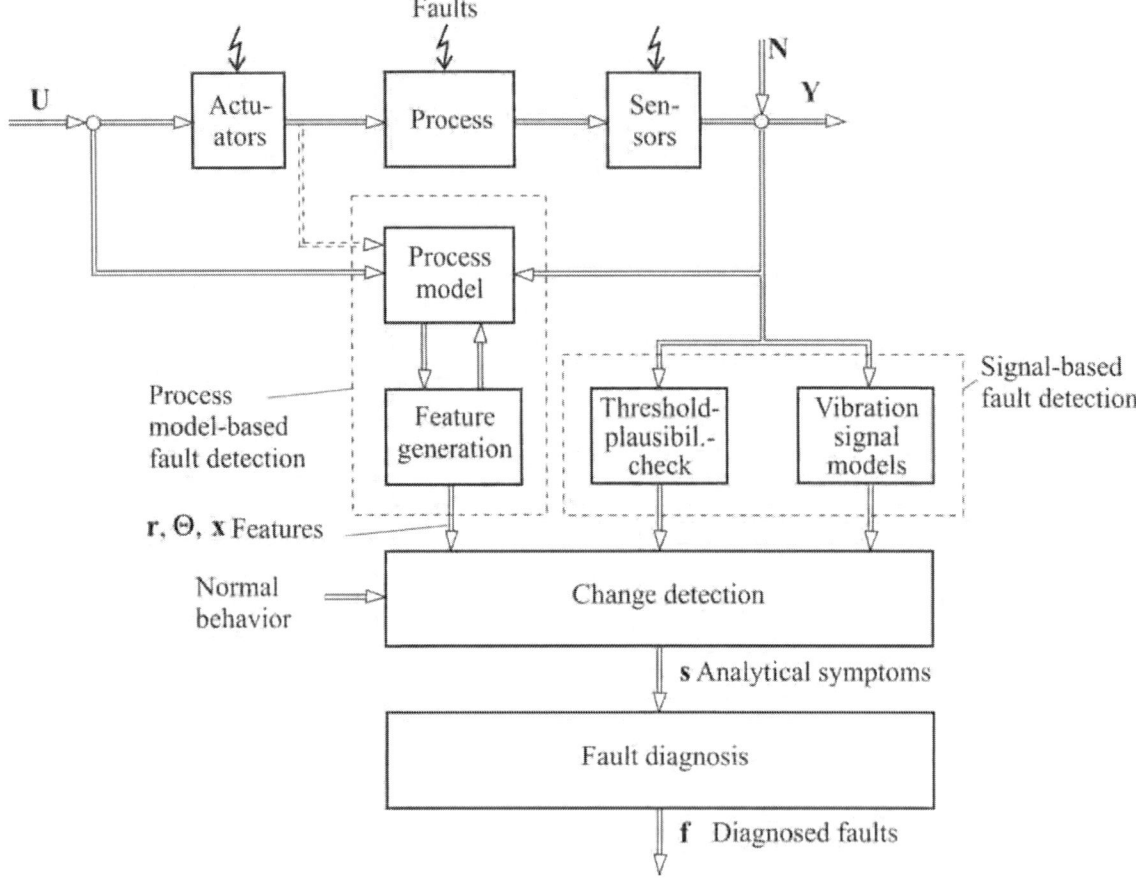

Figure 3. General scheme of process model-based and signal-based fault detection [4].

In order to obtain specific symptoms it is necessary to have more than one input and one output signal for *parity equations* or *output observers*. For *parameter estimation* one input and one output may be sufficient. Because of the various properties it is recommended to combine different methods in order to have a large fault detection coverage. A fault-detection capability usually is sufficient if used for fault-tolerant systems. Then, a fault diagnosis is not necessarily required. However, diagnosis capability is advantageous for general online supervision. Especially for on-board applications in automobiles, the allowable computations are very limited, which restricts the fault detection to methods with less computations on microcomputers. Furthermore, it is required that the fault-detection methods are transparent and easy to understand, function reliably for the different operating conditions, use only few measured signals and need only low effort for modeling. Also maintenance effort and easy transfer to modified components are important issues.

6 Fault-tolerant components

The discussion on high-integrity systems and automatic controlled systems shows that a comprehensive overall fault tolerance can be obtained by fault-tolerant components and fault-tolerant control. In the following some examples for fault-tolerant sensors and actuators are described.

6.1 Fault-tolerant sensors

A fault-tolerant sensor configuration should be at least fail-operational (FO) for one sensor fault. This can be obtained by applying *hardware redundancy* with the same type of sensors or by *analytical redundancy* with other sensors and process models.

a) Hardware sensor redundancy

Sensor systems with static redundancy are realized for example with a triplex system and a voter, Figure 1a). A configuration with dynamic redundancy needs at least two sensors and a fault detection for each sensor, Figure 1b). Usually, only hot standby is feasible. Another less powerful possibility is a plausibility check for two sensors, also by using signal models (e.g. variance), to select the more plausible one.

The fault detection can be performed by self-tests, e.g. by applying a known measurement value to the sensor. Another way are *self-validating sensors*, [16], where the sensor, transducer, and a microprocessor form an integrated, decentralized unit with self-diagnostic capability. The self-diagnosis takes place within the sensor or transducer and uses several internal measurements.

b) Analytical sensor redundancy

As a simple example a process with one input u and two measureable outputs y_1 and y_2 and the two processes G_1 and G_2 are considered, see Figure 4a). If the process models G_{M1} and G_{M2} are known, two redundant signal values $\hat{y}_1$ and $\hat{y}_{1u}$ can be calculated, if G_{M2} can be inverted. Based on the three signals y_1, $\hat{y}_1$ and $\hat{y}_{1u}$ a fault-tolerant signal $\hat{y}_{1FT}$ can be determined by a 2-out-of-3 selection and a voter. The same principle can be applied to obtain fault-tolerant signals $\hat{y}_{2FT}$ and $\hat{u}_{FT}$. An application of this sensor-fault-tolerance principle is shown in [17] for the position sensor of an electrical combustion engine throttle.

A similar scheme is depicted in Figure 4b). Three redundant signals are generated based on three measurements and three residuals r_1, r_2 and r_3 are determined by comparison with the original signals. A fault-residual table in Figure 4c) then shows unique patterns for the faults of all three sensors. Therefore, the detection of a respective sensor fault is possible and two fault-tolerant signals y_{1FT} and y_{2FT} can be determined.

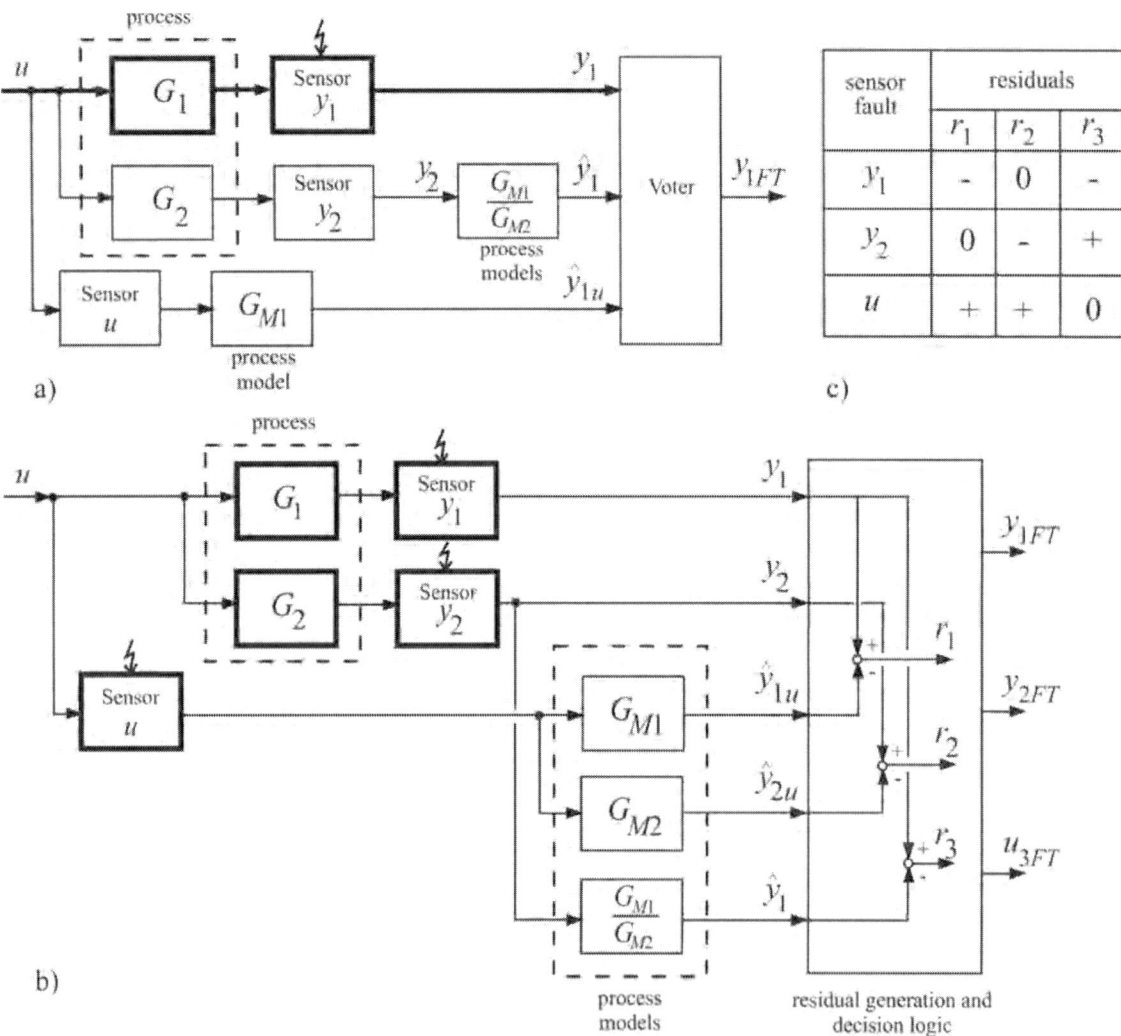

Figure 4. Analytical sensor fault tolerance for a process with one measured input u and two outputs y_1 and y_2: a) Two redundand values for y_1 are calculated based on measured u and y_2; b) Redundant values for y_1 and y_2 are residuals r_1, r_2 and r_3 are determind c) Fault-symptom table for b).

One example for this combined analytical redundancy is the yaw rate sensor for the ESC, where additionally the steering wheel angle as input is used to reconstruct the yaw rate through a vehicle model as in Figure 4b). Based on the lateral acceleration and the wheel speed difference of the right and left wheel (no slip) and vehicle models the yaw rate is reconstructed and given to a voter to form a fault-tolerant value as shown in Figure 4a), [18].

Figure 5 shows the principle of a fault-tolerant steering angle sensor [19]. The axle of the steering wheel is equipped with a gear wheel driving two pinions. Each pinion turns a permanent magnet whose position is sensed by one of two giant magneto resistances (GMR) measuring bridges. The two sensors use the nonius/vernier principle to obtain the absolute position over $\pm720°$. The evaluation of the sensors is integrated in two separate microcontrollers which supervise each other. In normal operation the master is connected to a bus and the slave monitors the master. In case of a fault, the master switches off and the slave is connected to the bus. Hence, the system is FO-FSIL. One fault of the sensor or microcontroller can be tolerated.

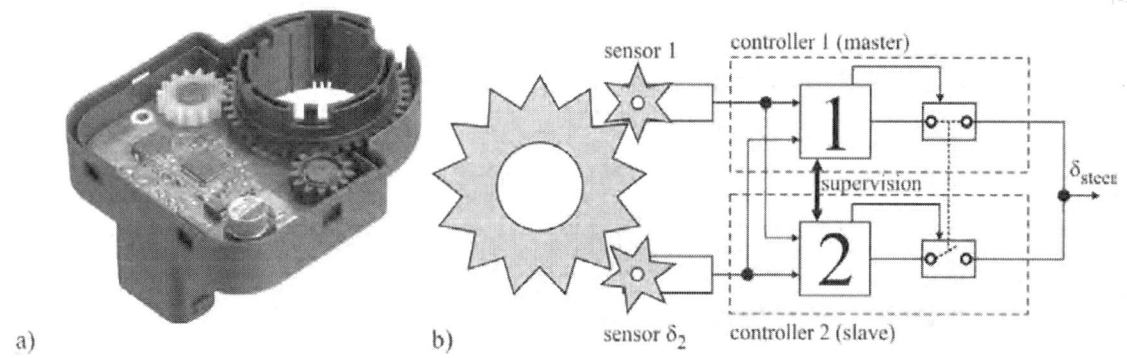

Figure 5. Fault-tolerant steering angle sensor [19]: a) System b) Schematic and signal flow.

6.2 Fault-tolerant actuators

Actuators generally consist of different parts: input transformer, actuation converter, actuation transformer and actuation element (e.g. a set of DC amplifier, DC motor, gear and valve, as shown in Figure 6a)). The actuation converter converts one energy (e.g. electrical or pneumatic) into another energy (e.g. mechanical or hydraulic). Available measurements are frequently the input signal U_i, the manipulated variable U_0 and an intermediate signal U_3.

Fault-tolerant actuators can be designed by using multiple complete actuators in parallel, either with static redundancy or dynamic redundancy with cold or hot standby (Figure 1) [20]. One example for static redundancy are hydraulic actuators for fly-by-wire aircraft where at least two independent actuators operate with two independent hydraulic energy circuits.

Another possibility is to limit the redundancy to parts of the actuator which have the lowest reliability. Figure 6b) shows a scheme where the actuation converter (motor) is split into separate parallel parts. Examples with static redundancy are two servo valves for hydraulic actuators, [21] or three windings of an electrical motor (including power electronics), [22].

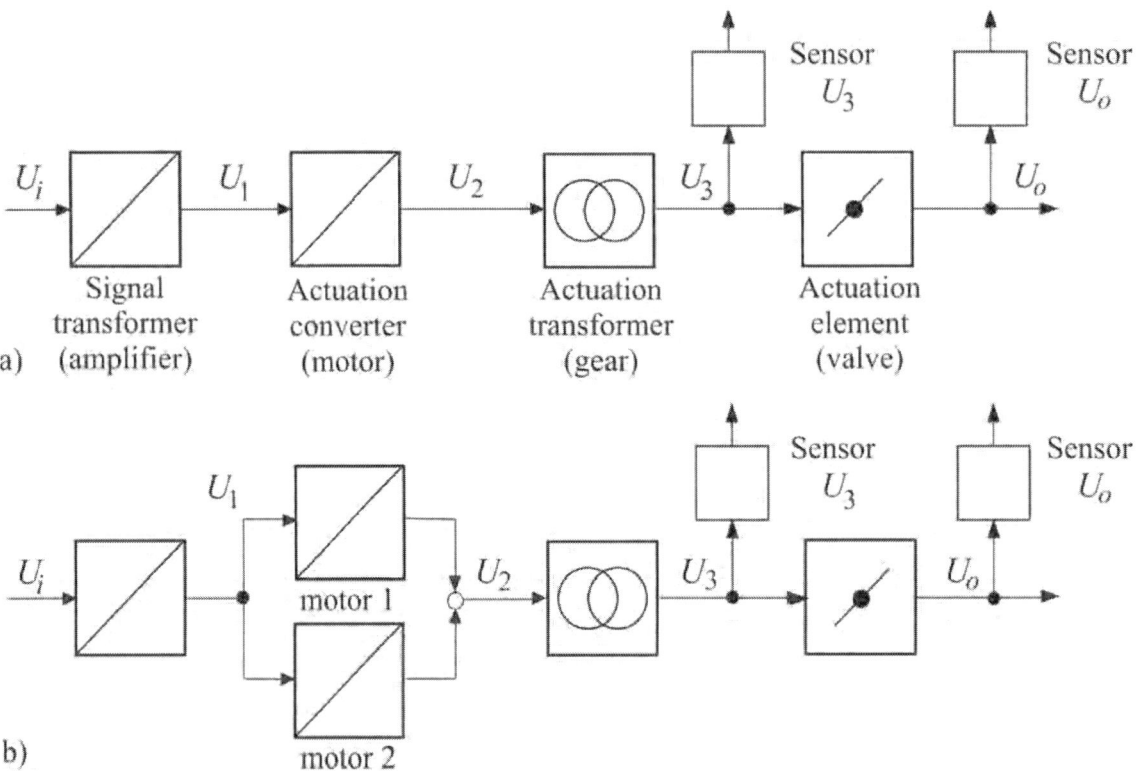

Figure 6. Fault-tolerant actuator: a) Common actuator. b) Actuator with duplex drive, two parallel motors and one gear. Hot or cold standby.

One example for dynamic redundancy with cold standby is the cabin pressure flap actuator in aircraft, where two independent DC motors exist and act on one planetary gear, [23], [24], [13].

As cost and weight generally are higher than for sensors, actuators with fail-operational duplex configuration are to be preferred. Then either static redundant structures, where both parts operate continuously, Figure 1a) or dynamic redundant structures with hot standby Figure 1b) or cold standby, Figure 1c) can be chosen. For dynamic redundancy fault-detection methods of the actuator parts are required. One goal should always be that the faulty part of the actuator fails silent, i.e. has no influence on the redundant parts.

a) Electrical duplex actuator system

Figure 7 depicts a *duplex actuator system* for the cabin pressure control in passenger aircraft. Two brushless DC motors (BLCD) act on a common gear. These DC motors and their microcontrollers operate alternatively form flight to flight. A model-based fault-detection system was developed to detect different faults of the DC motors [23], [24], [13]. In the case of a fault in one motor the other one is switched to be active. This system is therefore FO-F with dynamic redundancy and cold-standby.

A traditional redundant system for automobiles is the *hydraulic brake* designed as a mandatory dual-circuit transmission system with e.g. front-axle/rear-axle split or diagonal distribution pattern [25]. The braking pressure results form a combination of pedal force and (pneumatic) brake booster. The pedal force acts directly on the piston in a tandem master cylinder. Its pressure in a first chamber moves a floating intermediate piston and generates a pressure in the second chamber. Both chambers supply the respective circuits separately. Upon a fault (e.g. leak) in one circuit the other is still active, however, with reduced braking acceleration. Hence, the system is FO-F with dynamic redundancy and hot standby.

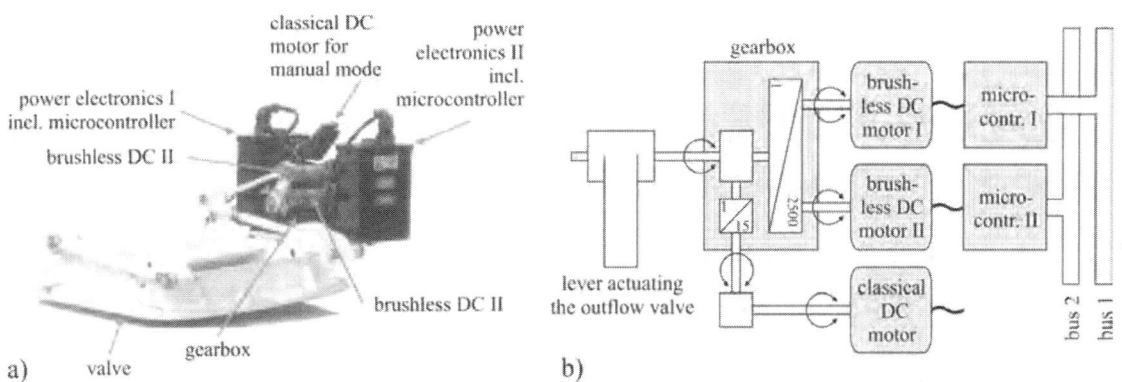

Figure 7. Fault-tolerant electrical actuator of the aircraft cabin outflow valve.
a) System. b) Schematic.

b) Fault-tolerant hydraulic actuators

Figure 8 shows the electro-hydraulic rudder actuator of the Eurofighter, see e.g. [26]. The actuator consists of a proportional-acting valve, which is driven by four separate solenoid units. Each one is supplied by its own power electronics. All solenoids are acting on the same valve spool. Two control edges each supply one cylinder chamber with hydraulic fluid. The cylinder has four chambers with identical active piston areas. The system can sustain leakages and/or pressure losses in one of the two hydraulic circuits. It can also sustain internal leakages in one of the two hydraulic circuits without losing the characteristic stiffness of hydraulic systems. As long as neither the valve spool nor the piston rod are affected, the system is FO-F.

A different prototypical realization of a *redundant hydraulic actuator* was developed by [27]. Here, only the valve has been doubled since it has been found by a detailed statistical analysis of maintenance records that the valve alone makes up roughly 51% of all faults at hydraulic servo axes. Upon the jam of one valve spool, the other valve can take over the volume flow. The entire servo axis is built from standard components, i.e. no design and construction of new, specialized components are necessary. This setup can tolerate all valve faults. As long as the piston rod is not affected and the spare valve is able to conduct any remaining parasitic flow of the faulty valve, the system is FO-F.

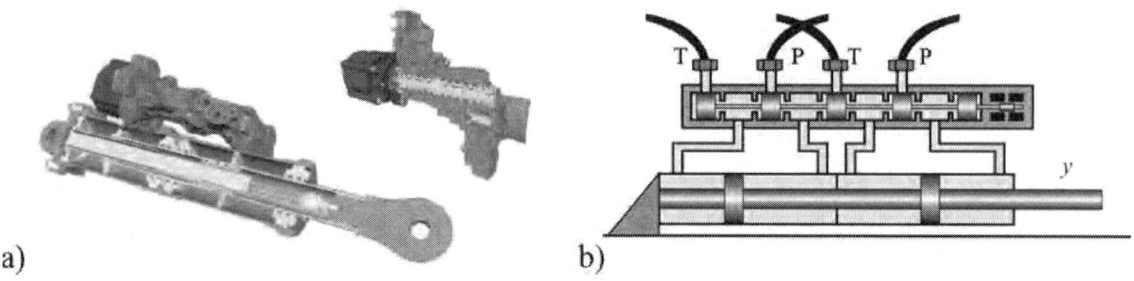

a) b)

Figure 8. Fault-tolerant electro hydraulic rudder actuator for the Eurofighter with two centralized hydraulic power supplies.

Another design with a dual-tandem ram, that is typical for aeronautical applications [28] is shown in Figure 9. This figure shows an actuation system for the F/A-18 horizontal stabilizer, which is a secondary control surface. In this example, the hydraulic cylinder is doubled and directly supplied with hydraulic fluid by two fixed-displacement pumps that are driven by two brushless DC motors. Bypass valves allow the piston to move even if the motor axle, respectively pump jams. The big advantage of this setup is that in the case of removal of the component, no hydraulic connections must be loosened. Furthermore, in the case of a leakage of the hydraulic piping/components, only the hydraulic oil of the component specific hydraulic circuit will spill, leaving all other hydraulic circuits of the plane unaffected. Provided that the valves are still functional and can disconnect the piston chambers from the pump, the system is FO-FSIL.

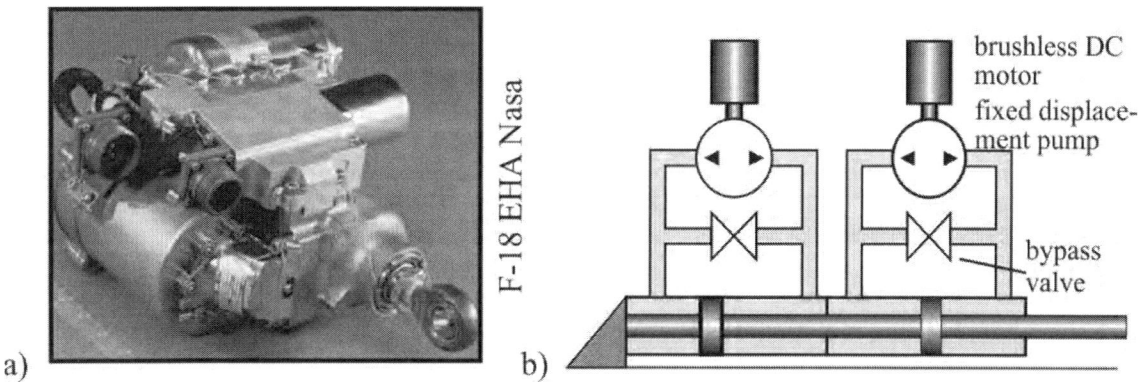

a)   b)

Figure 9. Fault-tolerant electro-hydraulic actuator with integrated, decentralized power supply for secondary flight control surface.

Many other architectures for electro-hydraulic actuators in aeronautical applications have been assessed in [29]. The big lead in X-by-wire functionalities and the accompanying use of fault-tolerant mechatronic components in aeronautical applications is typical and can be explained easily: Fly-by-wire functionalities for airplanes could be introduced with little increase in risk as control surfaces by themselves are redundant. Almost all maneuvers can be realized by different combinations of control surfaces.

6.3   Fault-tolerant electrical drives

a)   Fault-tolerant frequency converter

The *power electronics for frequency-controlled AC drives* have usually one inverter leg per motor phase. Upon the loss of one leg, the three-phase motor becomes single-phased and cannot generate a rotating magnetic field any longer. To overcome this fault situation, one can control each phase separately with a full H-bridge. If each phase of the motor is connected to a full H-bridge, the three-phase motor, upon the loss of one inverter leg or phase, can be operated as a two-phase motor and can still produce a rotating magnetic field. The major disadvantage is the fact that for separate H-bridges, each winding needs two wires to establish the connection to the power converter, making the wiring more expensive. For more details see [30]. The combination of a standard threephase PMSM (permanent-magnet synchronous motor) with an inverter with full H-bridges has been presented in [30], [31]. The system is now also FO-FSIL with respect to inverter or motor phase faults.

15

b) Multi-phase motors

An alternative to using full H-bridges for each phase is to design so called *multi-phase motors*, that have more than three phases. Although the first designs date back to the late 1960s, they have only recently been in the focus of research, as they are well suited for applications in the area of ship, locomotive, and electric/hybrid vehicle propulsion, see, e.g. [32], and the more electric aircraft, an initiative gathering momentum in the late 1990s with the aim to control aircraft subsystems with electrically actuated drives in place of mechanical, hydraulic or pneumatic means.

One general reason for the introduction of multi-phase motors in applications demanding the highest power is that semiconductors are not yet capable of switching the high currents that traditional three-phase motors would need to satisfy such high power demands. By increasing the number of phases, the power-per-switch can be limited to values that can be borne by the semiconductors. Also, multi-phase machines can easily sustain phase losses. Upon the loss of one phase, an n-phase motor becomes an $(n-1)$ phase motor $(n \geq 4)$ and hence shows a smaller loss in the power-rating and the uniformity of the circumferential torque distribution as the number of phases n increases. Furthermore, multi-phase motors show improvements in the noise characteristics and their torque production can easily be enhanced by the injection of higher-frequency harmonics.

The design of a four-phase fault-tolerant PMSM aircraft actuator is shown in [33]. A five-phase permanent magnet motor has been realized by [34] and has been investigated experimentally for post-fault operation. Depending on the number of windings, the topology of the windings and the allowable loss in torque, the system has a dynamic redundancy with hot standby and can sustain one or more faults. Thus, the system is at least FO-FSIL with respect to phase faults.

c) Duplex motors in parallel or serial connection

Instead of designing fault-tolerant components of electrical drives one may also take separate complete motors. This is especially of interest for bearing faults which may lead to a blockage of the rotor and the faulty motor can be detached by a clutch. As shown in Figure 10 either a parallel or serial configuration can be chosen [35], [20].

There are two load transfer modes possible for the *parallel connection*, Figure 10a), which are chosen depending on the severity of the fault, [35]: In the case of a *minor fault*, the second motor first powers up. Once the desired rotational velocity is reached, the first clutch disengages and thus separates the defective motor from the load and the second clutch engages and establishes the power flow between the spare motor and the load. In case of a *severe* fault, the first clutch disengages instantaneously. The load is

16

thus free-running while the second motor powers up and reaches the desired final velocity. Once the drive is up to speed, the second clutch closes and the load is driven by the second motor. As each motor can be decoupled from the load, it is also possible to conduct special motor tests from time to time without affecting the motion of the load.

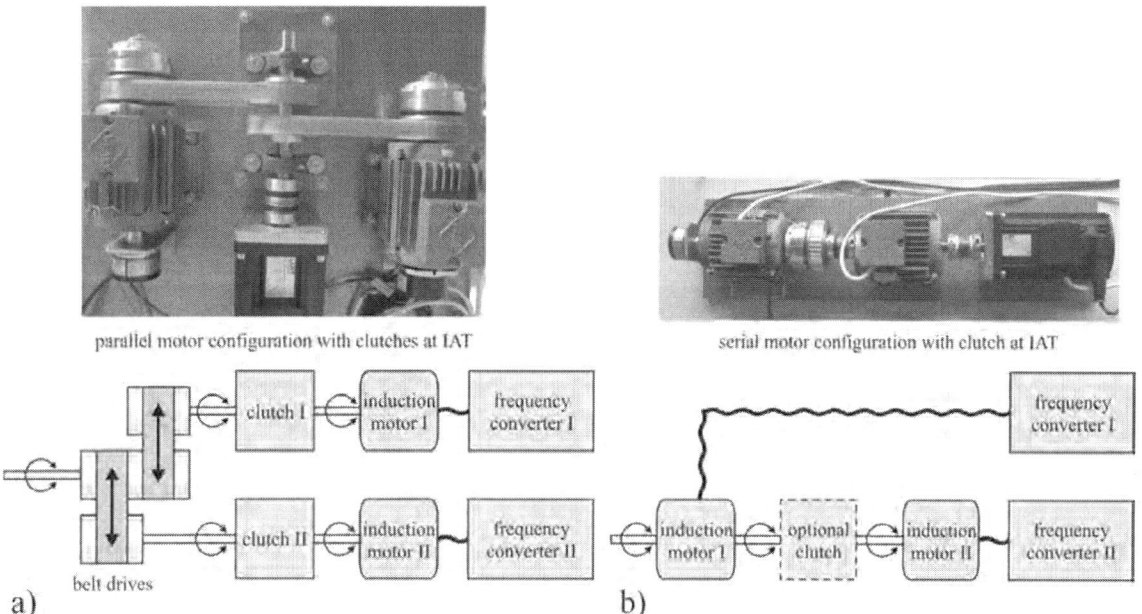

Figure 10. Fault-tolerant electrical drives. a) Parallel motor configuration. b) Serial motor configuration.

In the case of a *serial connection*, two motors are mounted one behind the other as illustrated in Figure 10b). Here, at least the rear motor can be detached from the load, while the front motor is always connected to the load. Thus, if the rotor of the front motor blocks, the entire fault-tolerant drive fails.

7 Fault-tolerant electrical power steering systems

For small and medium class passenger cars the hydraulic power steering system (HPS) was replaced by an electrical power steering since about 1996. One distinguishes column EPS (6–8 kN, 0.2–0.5 kW), pinion EPS (7–12 kN, 0.5–0.7 kW) and rack EPS (9–16 kN, 0.5–1 kW). Advantages are less installation effort, fuel saving, improved adaptation and the use of electrical input commands for driver assistance systems (automatic parking, lane keeping) and automatic driving. In the case of failures of the electrical power assisting system, the driver can take over the required steering torque for small and medium cars. However, for larger cars and light commercial vehicles the EPS should be fail-operational with regard to failures.

In the case of *automatic steering*, the driver is out of the loop. Automatic closed-loop systems cover usually smaller additive and parametric faults of the controlled process. However, if the faults become larger, either sluggish, less damped or unstable behavior may result. The not engaged driver may then not perform the right steering command in time. Therefore, it will be required that EPS systems in the case of automatic controlled steering must have *fail-operational functions* in the case of faults, i.e. have to be fault-tolerant.

Figure 11 illustrates a parallel-axial EPS and the available measurements. The field-oriented torque control of the permanently excited synchronous motor (PMSM) operates with reference values for the q- and d-current which are determined via the supporting steering torque characteristic $M_{ref}(M_{SC}, \varphi_{SC}, v)$ of the EPS, see the signal-flow diagram in Figure 12. The generated torque follows from

$$M_{el} = \frac{3}{2} n_p \left[\psi_{PM} i_q + (L_d - L_q) i_q i_d \right] \tag{1}$$

with n_p the pole pair number, ψ_{PM} the flux linkage, and i_q and i_d the torque and field generating currents. The resulting load torque of the mechanical part of the steering system is modelled by

$$M_{mech} = J_A \ddot{\varphi} + \left(d + \frac{d_{BS}}{i_{BG}^2} \right) \dot{\varphi} + \frac{c_{BS}}{i_{BG}^2} \varphi - \frac{d_{BS}}{i_{BG}^2} \dot{y} - \frac{c_{BS}}{i_{BG}^2} y \tag{2}$$

with J_A the ratio of inertia of the motor and the belt drive, φ the motor angle position, d damping coefficients, c_{BS} stiffness and i_{BG} gear ratio of the ball recirculation gear and the toothed belt gear, y the rack position [37–39].

Measurements:

- phase currents i_U, i_V, i_W
- manipulated voltages u_d, u_q
- duty cycles d_U, d_V, d_W
- rotor position φ_{PM}
- temperature (ECU) T_1
- steering wheel angle φ_{SC}
- steering column torque M_{SC}

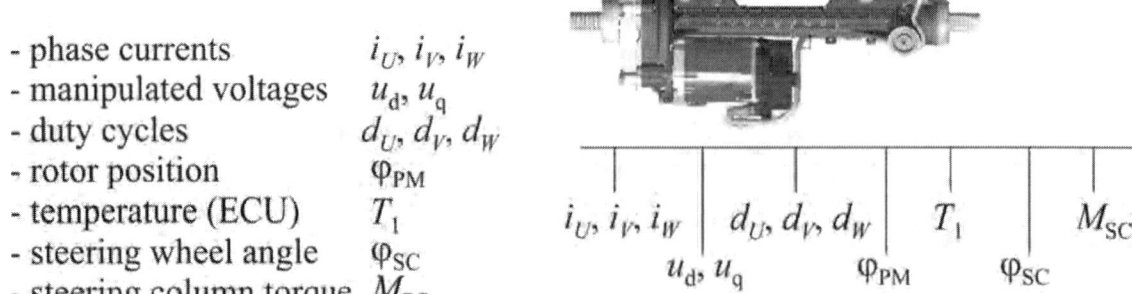

Figure 11. Parallel-axial actuator system for a rack EPS (ZF-Lenksysteme) and available measurements.

18

Based on these mathematical models parity equations can be stated [4], [13], and residuals between measured variables and outputs can be calculated which allow the detection of inherent faults [38].

For most components of the steering actuator system hardware redundancy is required in form of dynamic redundancy with hot or cold standby according to Figure 1b) and c). This means that a selection of components has to be doubled. Figure 13 depicts redundancy structures with different degrees of redundancy. At first the torque sensor can be duplicated or an analytical redundancy concept can be programmed [40]. In a next step the torque sensor, the ECU and inverter are duplicated, see Figure 13, case B. The arrangement can be cold or hot standby. In the case of a fault in one channel the other channel stays active and the faulty channel is switched off. Case C in Figure 13 has a further redundancy in the windings by a multi-phase configuration [41], [42]. A serial connection of two motors is shown in Figure 13, case D. Figure 13, case E despicts a duplication of the complete EPS actuator. In this parallel arrangement also the gear is duplicated [37], [38].

The degree of redundancy increases from case B to E, however on cost of hardware extent, installation space, cost and weight. The selection of the redundancies also depends on fault statistics for the different components. Case C seems to be a reasonable compromise, however requires a special motor design. In the case of a winding fault the power is reduced. In the cases D and E the power can be distributed differently to both motors in the normal operating range.

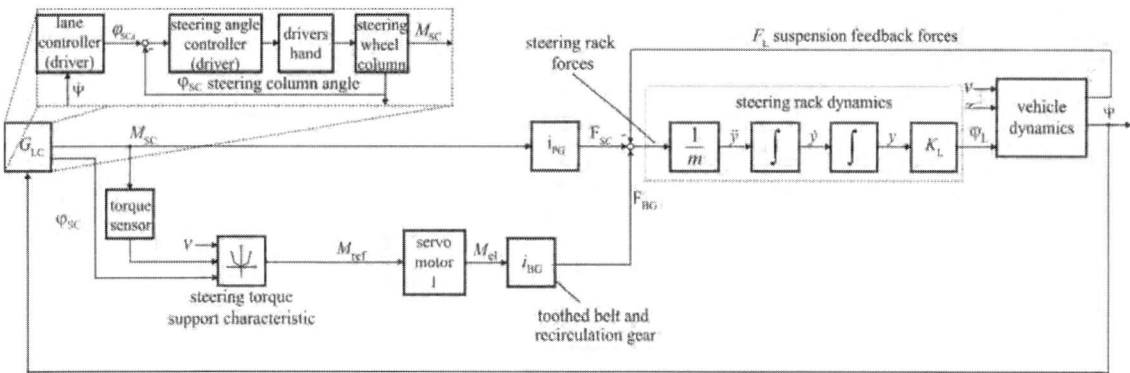

Figure 12. Signal-flow diagram of an electrical power steering system.

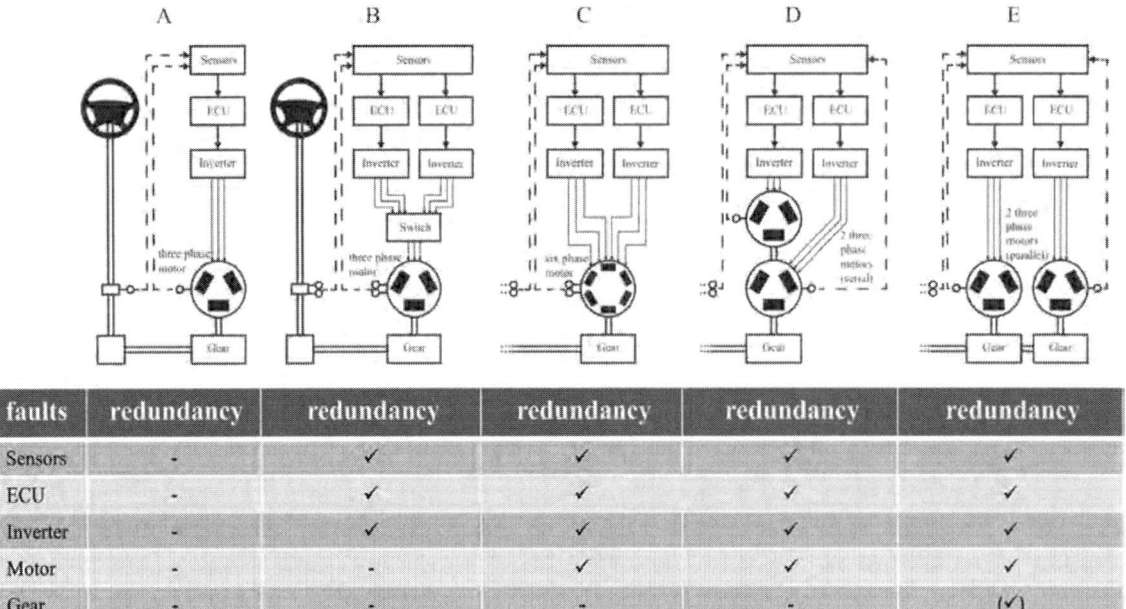

faults	redundancy	redundancy	redundancy	redundancy	redundancy
Sensors	-	✓	✓	✓	✓
ECU	-	✓	✓	✓	✓
Inverter	-	✓	✓	✓	✓
Motor	-	-	✓	✓	✓
Gear	-	-	-	-	(✓)

Figure 13. Redundancy structures for EPS systems.

Figure 14 depicts a block diagram for a fault-tolerant EPS system with two serial-connected servomotors. If two motors act on one mechanical connection as for cases D and E, a decoupling torque controller is required. A test bench with a duplex EPS system in parallel configuration by using two mass produced single EPS actuators is shown in Figure. 15. This test bench is used to investigate fault detection, control and fault-management methods [37], [39].

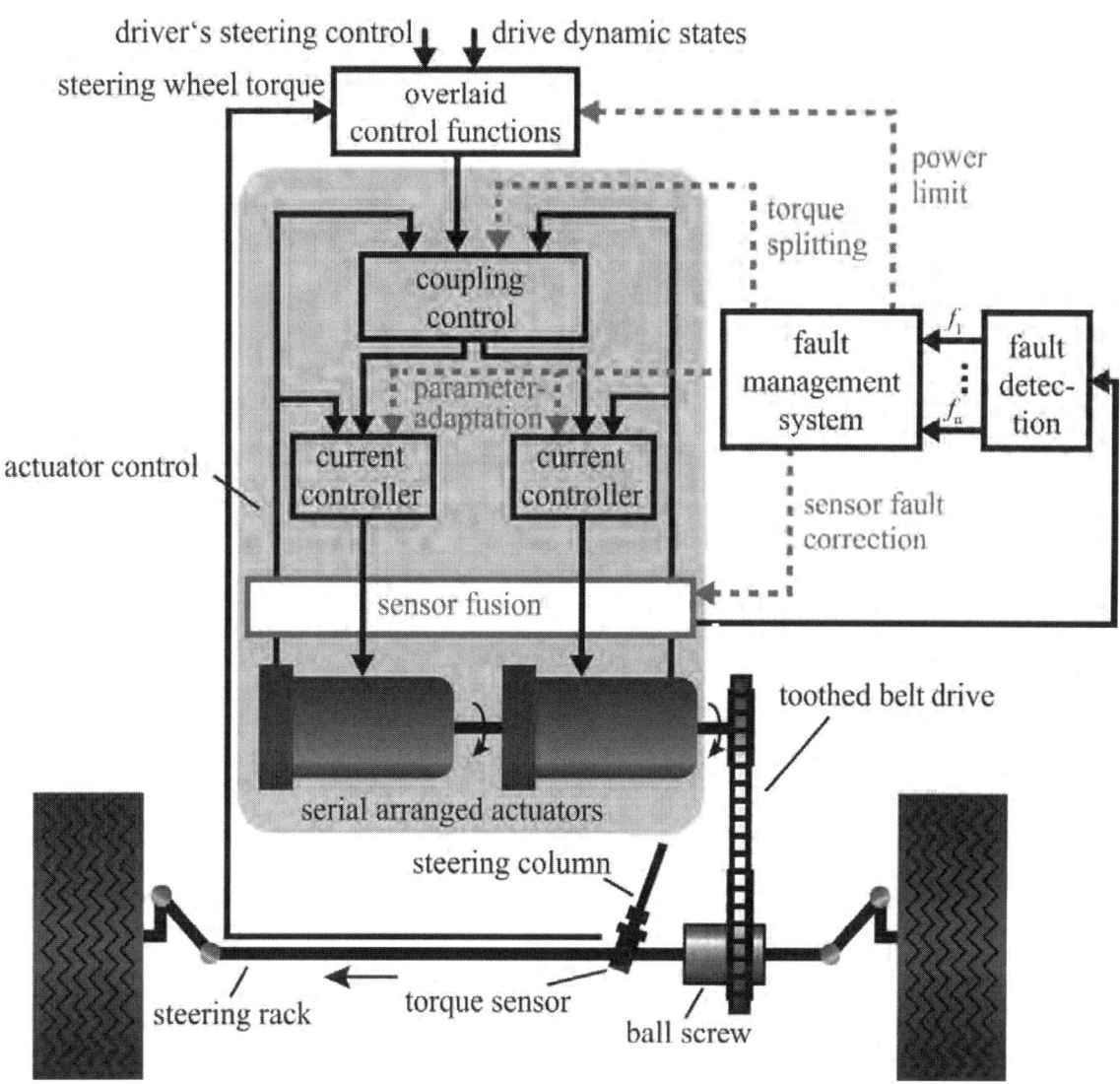

Figure 14. Fault-tolerant ESP system with a duplex servo motor in serial coupling, (case D).

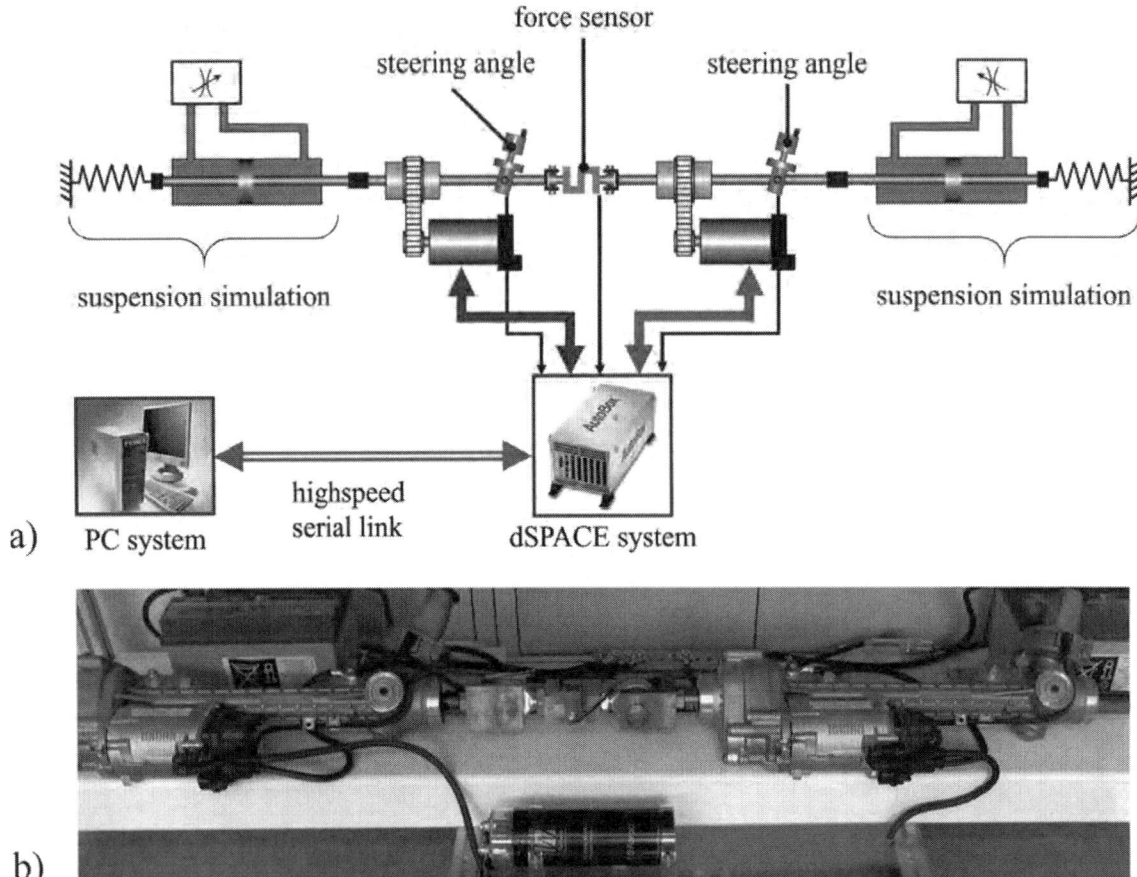

Figure 15. Testbench at IAT for a fault-tolerant duplex EPS system in parallel configuration, (case E). a) Schematic. b) Foto.

References

[1] IEC 26262, "Functional safety standard for automotive electrical/electronic systems", International Standard Organisation 2011.

[2] IEC 61508, "Standard draft IEC 61508, Part 1-7", Functional Safety of E/ E/ PES: (complex) Electrical/ (complex) Electronic/ Programmable Electronic Systems. Version 4.0, 1997.

[3] N. Storey, "Safety-critical computer systems", Essex, UK: Addison Wesley Longman Ltd., 1996.

[4] R. Isermann, "Fault-diagnosis systems", Heidelberg: Springer, 2006.

[5] R. Isermann, R. Schwarz, S. Stölzl, "Fault-tolerant drive-by-wire systems – concepts and realizations", IEEE Control Systems Magazine, vol. 22, no. 5, pp. 64-81.

[6] N. Leveson, "Safeware. System safety and computer", Reading, MA, USA: Addison-Wesley Publishing Company, 1995.

[7] W.-D. Jonner, H. Winner, L. Dreilich and E. Schunck, "Electrohydraulic brake system – the first approach to brake-by-wire technology" SAE Technical Paper Series, no. 960991. In [4], pp. 221-228.

[8] H. E. Rauch, "Autonomous control reconfiguration", IEEE Control Systems Magazine, vol. 15, no. 6, pp. 37-48, 1995.

[9] P. R. Chandler, "Reconfigurable flight control at Wright laboratory", Neuilly-sur-Seine, France, vol. III, AGARD advisory report 360, Aerospace 2020, 1997.

[10] R. J. Patton, "Fault-tolerant control: the 1997 situation", in IFAC Symposium on Fault Detection, Supervision and Safety for Technical Processes (SAFEPROCESS), Kingston Upon Hull, UK, vol. 2, pp. 1033-1055, 1997.

[11] J. Chen, R. J. Patton and Z. Chen, "Active fault-tolerant flight control systems design using the linear matrix inequality method", in Transactions Institute of Measurement and Control, vol. 21, no. 2/3, pp. 77-84, 1999.

[12] S. Suryanaryanan and M. Tomizuka, "Fault-tolerant lateral control of automated vehicles based on simultaneous stabilization", in 1st IFAC Conference on Mechatronic Systems, Darmstadt, Germany: 2000.

[13] R. Isermann, "Fault-diagnosis applications", Heidelberg: Springer, 2011.

[14] J. J. Gertler, "Fault detection and diagnosis on engineering systems", New York, NY, USA,:Marcel Dekker, 1999.

[15] J. Chen and R. J. Patton, "Robust model-based fault diagnosis for dynamic systems", Boston, MA, USA: Kluwer Academic Publishers, 1999.

[16] M. P. Henry and D. W. Clarke, "The self-validating sensor: rationale, definitions, and examples", in Control Engineering Practice, vol. 1, no. 2, pp. 585-610, 1993.

[17] T. Pfeufer, "Model-based fault detection and diagnosis for an automotive actuator", (in German), Fortschr.-Ber. VDI Reihe 8 No. 764. Düsseldorf, Germany: VDI-Verlag, 1999.

[18] A. T. van Zanten, R. Erhardt, K. Landesfeind and G. Pfaff, "VDC systems development and perspective", SAE Technical Paper Series, no. 980235. In [4], pp. 373-394.

[19] S. Quass and P. Schiebel, "Aspects of future steering markets and their relevance to steering sensors", Proceedings of IQPC – Advanced Steering Systems, 2nd Annual Conference, May 21-23, Frankfurt, 2007.

[20] M. Münchhof, M. Beck, R. Isermann, "Fault-tolerant actuators and drives – structures, fault-detection principles and applications", Annual Reviews in Control, vol. 33, pp. 136-148, 2009.

[21] R. Oehler, A. Schoenhoff and M. Schreiber, "Online model-based fault detection and diagnosis for a smart aircraft actuator" in IFAC Symposium on Fault Detection, Super-vision and Safety for Technical Processes (SAFEPROCESS), Kingston upon Hull, UK, vol. 2, pp. 591-596, 1997.

[22] A. Krautstrunk and P. Mutschler, "Remedial strategy for a permanent magnet synchronous motor drive", Proceedings of EPE´99, Lausanne, Switzerland, 1999.

[23] O. Moseler, T. Heller, R. Isermann, "Model-based fault detection for an actuator driven by a brushless DC motor", Proceedings of the 14th IFAC World Congress, Beijing, China, vol. P, pp. 193-198, 1999.

[24] O. Moseler, "Mikrocontrollerbasierte Fehlererkennung für mechatronische Komponenten am Beispiel eines elektromechanischen Stellantriebs", Fortschr.-Ber. VDI Reihe 8, 980. VDI Verlag: Düsseldorf, 2001.

[25] Robert Bosch GmbH (ed.), "Automotive Handbook", 8th edition. Cambridge: Bentley publishers, 2011.

[26] R. Kress, "Robuste Fehlerdiagnoseverfahren zur Wartung und Serienabnahme elektrohydraulische Aktuatoren", Doctoral thesis. Technische Universität Darmstadt, Fachbereich Maschinenbau, 2002.

[27] M. Münchhof, "Model-based fault detection for a hydraulic servo axis", Dissertation Technische Universität Darmstadt. Fortschr.-Ber. VDI Reihe 8, 1105. Düssedorf: VDI Verlag, 2006.

[28] Moog Aircraft Group, "Redundant Electrohydrostatic Actuation System - Application: F/A-18 C/D Horizontal Stabilizer", 1996.

[29] T. Sadeghi and A. Lyons, "Fault tolerant EHA architectures", IEEE Aerospace and Electronic Systems Magazine, 7(3): pp. 32–42, 1992.

[30] S. Green, D.J. Atkinson, B.C. Mecrow, A.G. Jack and B. Green, "Fault tolerant, variable frequency, unity power factor converters for safety critical PM drives", IEE Proceedings – Electric Power Applications, 150(6): pp. 663–672, 2003.

[31] A. Krautstrunk, "Fehlertolerantes Aktorkonzept für sicherheitsrelevante Anwendungen", Aachen, Germany: Shaker Verlag, 2005.

[32] E. Levi, "Multiphase electric machines for variable-speed applications", IEEE Transactions on Industrial Electronics, 55(5): pp. 1893–1909, 2008.

[33] G.J. Atkinson, B.C. Mecrow, A.G. Jack, D.J. Atkinson, P. Sangha, and M. Benarous, "The design of fault tolerant machines for aerospace applications", Proc. IEEE International Conference on Electric Machines and Drives, pp. 1863-1869, 2005.

[34] N. Bianchi, S. Bolognani and M.D. Pre, "Impact of stator winding of a five-phase permanent-magnet motor on postfault operations", IEEE Transactions on Industrial Electronics, vol. 55, no. 5, pp. 1978-1987, 2008.

[35] J. Reuss and R. Isermann, "Umschaltstrategien eines redundaten Asynchronmotoren-Antriebssystems", SPS/IPC/DRIVES 2004: Elektrische Automatisierung, Systeme und Komponenten: Fachmesse and Kongress, Nürnberg, Germany, pp. 469-477, 2004.

[36] R. Reichel, "Steuersysteme im Flugzeug, fly-by-wire", Automatisierungstechnik, 52(12): pp. 588-595, 2004.

[37] M. Beck and R. Isermann, "Modelling of a duplex electrical power steering prototype", IQPC Conference on Steering Systems, Wiesbaden, Germany, November 2010.

[38] P. Kessler, "Model-based fault diagnosis of an EPS system", IQPC Conference on Steering Systems, Düsseldorf, Germany, 23-24 November 2015.

[39] R. Isermann and M. Beck, "Modellbasierte Methoden zur Erhöhung der Verfügbarkeit und Sicherheit von Fahrwerkomponenten", VDI/VDE-Tagung AUTOREG, Baden-Baden, Germany, 2011.

[40] F. Schöttler, "Functional safety in electrical power steering systems", IQPC Conference on Steering Systems, Frankfurt, Germany, 11-14 November 2013.

[41] J. Hayashi, "Road map of the motor for an electric power steering system", 4. ATZ-Konferenz chassis. tech plus, München, Germany, 2013.

[42] S. Yoneki, E. Hirozumi and B. Collerais, "Fail-operational EPS by distributed architecture", 5. ATZ-Konferenz chassis. tech plus, pp. 421-442, München, Germany, 2014.

[43] H.D. Heitzer, "Development of a fault-tolerant steer-by-wire system", Auto Technology, vol. 4, pp. 56-60, 2003.

[44] M. Hell, "Steer-by-wire: from concepts to reality", IQPC Conference on Steering Systems, Düsseldorf, Germany, 23-24 November 2015.

Deterministic architecture and middleware for domain control units and simplified integration process applied to ADAS

Dr. Georg Niedrist, TTTech Automotive GmbH

© Springer Fachmedien Wiesbaden GmbH, ein Teil von Springer Nature 2018
R. Isermann (Hrsg.), *Fahrerassistenzsysteme 2016*, Proceedings,
https://doi.org/10.1007/978-3-658-21444-9_15

Introduction

Modern cars offer an ever increasing number of electronic functions in all vehicle domains, like Advanced Driver Assistance Systems (ADAS), infotainment systems, vehicle dynamics systems and hybrid and electric drivetrains. In the past, each new customer function required yet another electronic control unit (ECU); due to cost, packaging, wiring and thermal constraints, this is not a suitable solution anymore. Today and even more so in the future we observe a trend towards modularization of automotive electronic systems and implementation of new customer functions by software only, effectively reducing the number of ECUs or at least not increasing it further.

In parallel to those technically mandated changes, we can also observe a change of supply chain and cooperation models: instead of sourcing a complete closed-box system including sensors, ECUs and possibly actuators from a single Tier1 supplier, OEMs increasingly turn to a more open "cherry picking" model, i.e. selecting the best-in-class elements (sensors and sensor processing SW, ECU HW and platform SW, application SW modules, actuators) separately from a number of Tier1 and SW suppliers and integrating the complete system on their own or in a network of partner companies. This approach is driven by the OEMs' needs for differentiation, which also results in increasing in-house development of customer functions by the OEMs.

All of those trends lead to the emergence of central domain ECUs based on platform architectures. For example, in the ADAS domain the requirement for central sensor fusion quite naturally leads to a central fusion and application controller architecture. Generic platforms are needed to foster function SW reuse across different car models, which in turn is necessary to cope with the immense costs and efforts for function validation especially in the ADAS domain.

Due to a changed cooperation model and the dramatically increased technical complexity, in this scenario not just the new role of the "SW integrator" emerges, but also the requirements on the domain ECU architecture, the SW platform and the integration process are very different from traditional ECUs. In the following sections we discuss the requirements on such domain ECU architectures with a particular focus on ADAS systems and present a novel solution which is currently employed in full series development with a variety of partners.

Requirements for ADAS domain ECUs

Figure 1 gives an abstract view of a high-end ADAS system like for instance needed in autonomous driving scenarios.

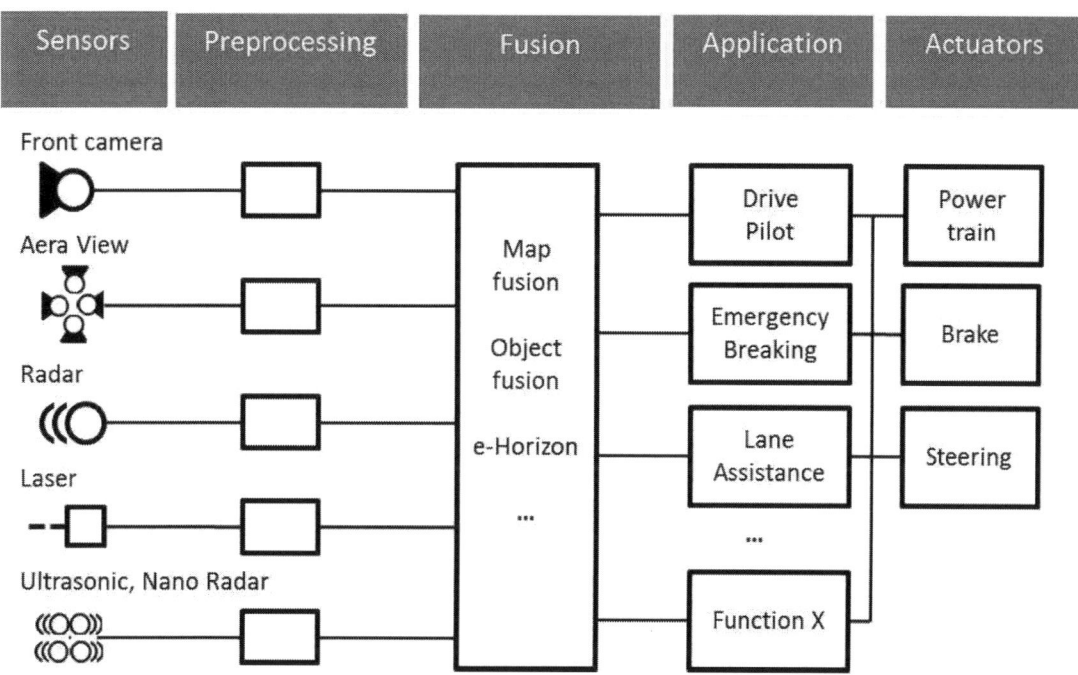

Figure 1: Abstract view of high-end ADAS system

A set of diverse sensory inputs like radar, cameras, lidar and ultrasonic sensors are processed to identify elementary pieces of information about the car's environment (lanes, traffic signs, people, other cars, ...), which are fed into a central fusion layer to compute a complete and consistent representation of the car's surrounding and the trajectories of all objects. The application layer builds upon that fused information, devises the driving strategy and implements the navigation and control algorithms to drive the vehicle, i.e. controls steering, braking and the drivetrain. Besides this core function of autonomous driving control, a variety of additional comfort and utility functions will usually be hosted by a central ADAS controller – from customer functions like Automated Emergency Braking (AEB), lane assistance and surround view to vehicle utility functions like event logging and black-box data recording functionalities.

The SW functions hosted by a central controller will depend on the OEMs overall vehicle architecture, e.g. it might include raw data sensor processing or alternatively rely on object data received from "smart sensors" attached to the central controller. In any case, at least the sensor fusion layer plus all the application and utility functions need to be integrated. Since the complexity of the task and most architecture considerations are relevant also for this "simpler" case, we will not elaborate on the concrete system architecture further.

The requirements on the HW and SW architecture of an ADAS domain controller come from many sources – performance, safety, the application SW development and integration process, mastering the technical complexity, as well as modularity and scalability

3

needs in order to support different vehicle options. Similar requirements have been present in the aerospace industry for a long time, leading to the well-established architecture of (Distributed) Integrated Modular Avionics ((D)IMA). The approach presented here is thus derived from DIMA and adapted to the specific requirements of the automotive domain in terms of cost, functionality and scalability.

- Performance: The highest possible processing power is a natural requirement in the ADAS domain, given the immense computational demand of sensor processing, sensor fusion and vehicle control. Depending on the overall system architecture, image processing devices, e.g. GPUs, might be needed, alongside high-performance general purpose processing devices like multicore System-on-Chip designs (SoCs) designs, e.g. based on ARM A53 or A57 cores. Devices from the consumer and infotainment world are on the forefront with respect to high-integration and performance/power ratio and need to be included.

- Safety: ADAS domain controllers will need to conform to ISO262626 ASIL C/D to support autonomous driving functions. This overall safety requirement currently cannot be met by high-end SoCs, and is therefore usually broken down by the means of ASIL decomposition[1] and clever SW partitioning. For this reason, an ADAS domain controller will usually incorporate at least one ASIL D capable automotive microcontroller in parallel to high-end SoC devices with lesser safety capabilities. Safety capabilities on the HW side need to be supplemented by platform SW mechanisms (memory partitioning, timing supervision, protection of communication between applications, HW diagnostics and supervision) to give full support for the applications.

- Interfaces: ADAS domain controllers usually require a set of traditional automotive network interfaces (CAN, FlexRay, LIN) in addition to Ethernet and possibly raw data video interfaces (LVDS, CSI) for connection of sensors. Note that along with CAN or FlexRay usually utility functions like diagnostics, calibration and security features need to be supported, which can be addressed properly by using an AUTOSAR Basic SW. High-end SoCs might not offer those interfaces and services and need to be supplemented by a traditional automotive microcontroller.

- ECU Modularity and Scalability: To support vehicle options, a domain controller will need to come in several variants from a top-line system to basic functions. SW reuse across all variants is mandatory, in order not to end up in developing separate ECU SW packets for each variant. Abstraction of the underlying HW, operating sys-

---

1 ASIL decomposition means that a higher safety integrity level (ASIL) can be reached by combining elements with lower levels in an appropriate way; for example two elements which each fulfil ASIL B requirements can result in an ASIL C or D capable system.

tems and communication mechanisms is essential – the application SW components (SWCs) shall be completely independent of the underlying platform implementation, which shall not contain any ad hoc application specific elements. Under this condition we can add or omit processing elements as needed by performance and safety demands for the different ECU variants, and still freely reuse and move SWCs between processors, decreasing SW developments and validation efforts drastically.

- Development Process: Application SW in the ADAS domain is often developed initially in a rapid prototyping scenario using modelling tools like ADTF or MATLAB/Simulink. It is highly desirable to support a seamless transition from such PC-based development to an industrialized implementation on the target ECU by ensuring continuous usability of test cases across all phases and allowing ideally the identical code to be run on PC or on the target. Mixed configurations – PC-based SWCs integrated with target-based SWCs – in a Software-in-the-loop (SIL) setup shall be possible.

- SW Integration Process: the HW and SW platform architecture shall allow the controlled integration of large number of SWCs coming from possible many different teams. "Controlled" means that the SW integrator (be it OEM or 3rd party) must not be overloaded with debugging, mediation and conflict resolution tasks among SWC suppliers: we want to avoid a scenario where single SWCs work just fine, but the completely integrated system does not because of timing interaction and resource conflicts, leading possibly to an endless procedure of iterations and incremental fixes. So we need *composability*, i.e. individually tested SWCs shall work immediately when integrated together. This also means that predictability with regard to resource consumption, runtime, data flow latencies and sequences of SWCs are key requirements. Finally, a black-box integration approach (no source code, anonymized interfaces) is desirable to support IP protection among potentially competing SWC suppliers.

- Mastering Complexity: a high-end central ADAS controller might approach the complexity of a complete vehicle electronics just a few years ago. A clean SW architecture and the abstraction of HW and operating systems shall provide the required modularity, portability and simplicity of interfaces for the SWCs. Observability of all ECU-internal communication through appropriate debugging utilities (of course without influencing system timing, like available on vehicle level through data logging and tracing), and the above mentioned requirements on the SW integration properties without mutual side effects are not just a matter of comfort, but a question of feasibility.

All of these requirements are not just valid for an "open" cooperation scenario where an OEM specifies, sources and integrates the complete system from contributions of variety of suppliers, but also in the more "closed" environment of a single Tier1 system supplier integrating a complex domain ECU with a multitude of in-house teams and possibly OEM contributions.

HW and SW Architecture Overview

Figure 2 shows an abstract block diagram of the ECU HW architecture.

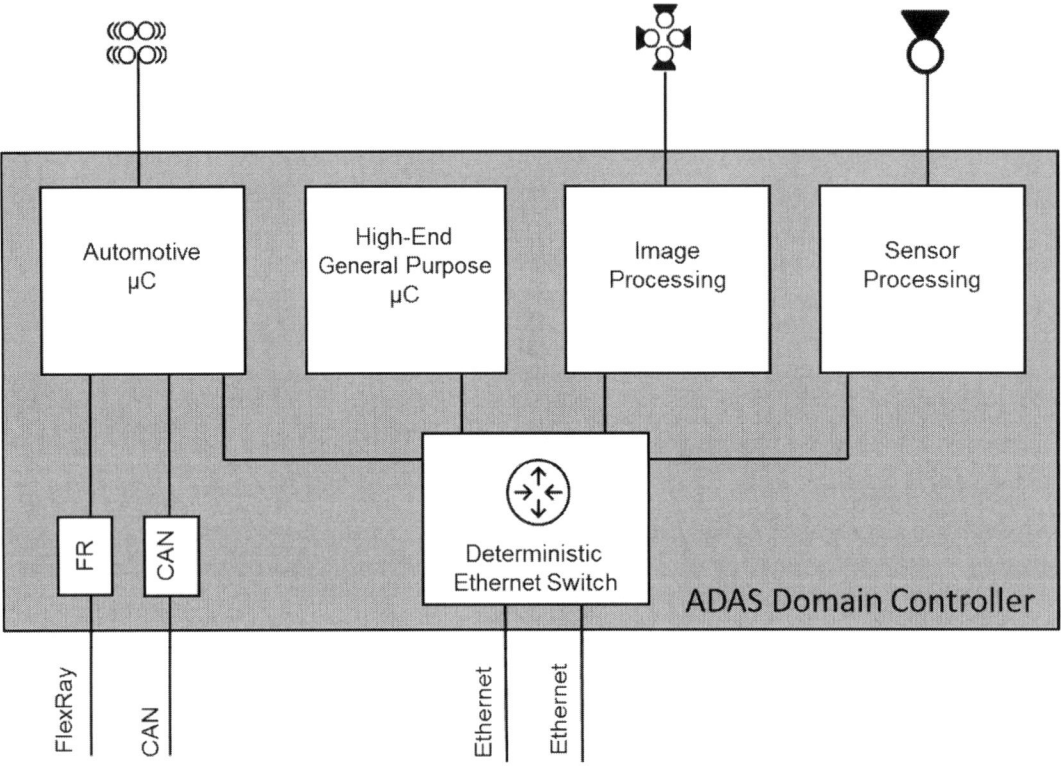

Fig. 2 Block Diagram of ECU HW Architecture

As indicated above, the architecture consists of a diverse set of processing elements: an automotive microcontroller (like Infineon Aurix or Renesas RH850), a general purpose processing engine based on an FPGA (like Altera Cyclone V), an image processing device (like devices from Nvidia or Renesas R-Car H3) and further specialized sensory processing devices. All devices are interconnected by a Deterministic Ethernet switch, and there are no direct point-to-point connections which hinder debugging and tracing of intra-ECU communication – which would clearly be a showstopper for integration and debugging of the highly complex interaction between a multitude of SWCs[2]. External interfaces are provided by CAN and FlexRay to the traditional car network, and by Ethernet for connection to high-end car networks and as a debugging interface.

The Deterministic Ethernet (DE) switch can be integrated in an FPGA or included as a standard off-the-shelf device like NXP's SJA1105T; besides the normal best effort Ethernet switching capability, it provides a *time-triggered*, deterministic mode of operation where all communication and switching takes place according to a predefined schedule in a completely predictable way (similar to FlexRay's mode of communication). All processing elements are synchronized to the switch and schedule their internal task processing accordingly; thus it provides the central heartbeat of the ECU and ensures deterministic, collision and jitter free communication among the SWCs on the processing hosts. The DE switch thus is the core element to ensure the above mentioned composability and deterministic integration property. Note that the processing elements just use their standard, integrated Ethernet MAC devices and need no special HW precautions (there are, however, SW enhancements on driver and OS level to support synchronized communication).

Safety is implemented by ASIL decomposition across devices, with the Automotive microcontroller providing the ASIL D capability, the FPGA implementing up to ASIL B measures, and the GPU being used for QM-rated algorithms. All devices offer memory protection and the associated Freedom-from-Interference to support mixed-criticality operation. End-to-end communication protection is applied for all safety-relevant inter-SWC (intra-ECU) communication.

Figure 3 gives a high-level SW architecture overview.

---

2 A possible exception might be direct, specialized point-to-point connections between devices to share raw sensory data. However, this has no impact on the communication and integration of SWCs and is therefore not further treated here.

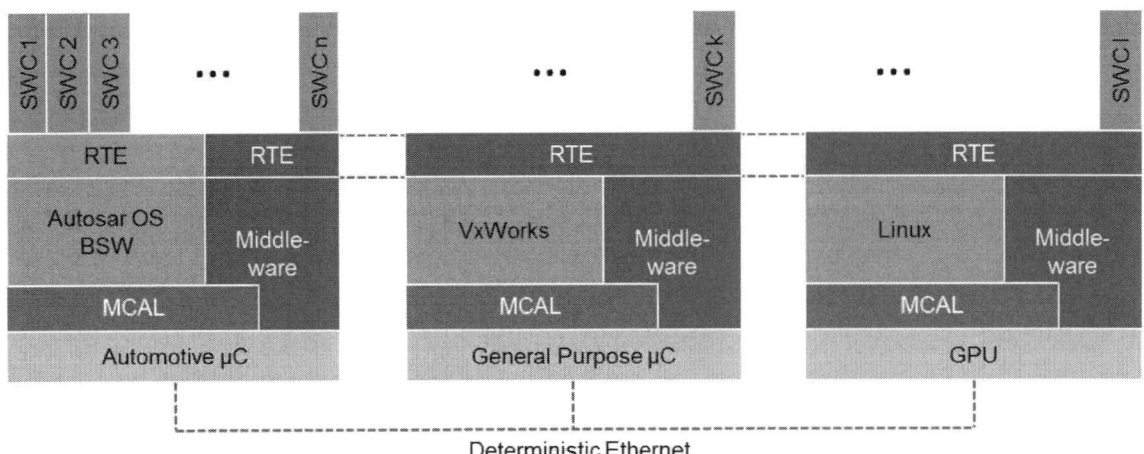

Fig. 3: SW Architecture

Operating systems are selected according to a best-fit approach:

- AUTOSAR Basic SW and OS are used on the automotive microcontroller. This contains the proven SW stacks for vehicle communication interfaces and basic SW services, as well as commonly used automotive functions (diagnostics, calibration). This choice ensures compliance to OEM vehicle networks and operation requirements, and also allows easy integration of OEM specific utility and legacy functions like special in-house security protocols.

- Wind River's VxWorks is used as a safety-enabled OS for the general-purpose SoC (FPGA), where safety-relevant applications up to ASIL B need to be run, but their performance demands exceed the automotive microcontroller's specifications. Besides the safety capability, VxWorks offers a wealth of proven functionalities and services, and POSIX compliance.

- Linux is the natural choice for non-safety-related processing elements on high-end SoCs (GPUs), in particular since it offers easy, often out-of-the-box integration of graphics processing libraries like OpenCL, OpenCV, CUDA and similar services.

This selection is quite natural taking the different processing elements into account, but leads to a very heterogeneous OS solution – without further measures the applications will find a completely different operating environment and API on each host, which strongly ties each application SW component to a predefined host. Additionally, basic mechanisms like communication and data management are not compatible on the different OSs: without further precautions the applications would need to implement proprietary ad hoc mechanisms and adaption layers, very likely mashing up platform functionalities with application code.

To provide a common execution environment and API on all hosts, and to ensure HW and OS abstraction, our solution is to place a generic middleware including an AU-TOSAR RTE type of communication layer on top of each OS. This RTE layer plus a broad set of utility functions form the generic middleware on all hosts which offers one common, host-independent API to the SWCs and achieves full HW and OS abstraction, with complete location transparency, i.e. all services are available on all hosts transparently. This ensures full portability of SWCs, i.e. they may be located flexibly to processing elements according to their requirements on performance, memory and safety. It also allows to move SWCs freely between processing cores and hosts should those requirements change or to optimize the utilization of processing power and memory of the ECU. For the SW integrator this amounts simply to a configuration step at compile time, without any code change of the SWC!

To achieve the deterministic integration property with its benefits (predictability, no integration side effects), the time-triggered mode of operation is implemented also in the middleware, by means of OS specific enhancements to synchronization and task scheduling/dispatching.

Finally, the middleware layer is also extended to the PC environment, allowing seamless SIL interaction (co-simulation) among applications on the PC and the target ECU and bridging the gap between rapid-prototyping and industrialized SW components – all while keeping all timing relations fully intact, and without any necessary code change from part of the SWCs.

To sum up there are two key elements used, in order to fulfil all requirements on performance, safety, the development and integration process, as well as modularity and scalability:

- A deterministic architecture exploiting the time-triggered paradigm for (ECU internal) network communication and task scheduling

- A common, generic middleware layer abstracting the diversity of the underlying microcontrollers and OSs

It should be stressed that this approach is completely scalable, as processing hosts may be added for different ECUs variants or car models without impact to the overall architecture and allowing complete reuse of SWCs across those variants. The same architecture can even be used in a distributed fashion on a vehicle level, e.g. by using OABR Ethernet interfaces between ECUs. Also it is remarkable that all core elements build upon well-established and standardized technologies like AUTOSAR, POSIX and time-triggered communication as defined in SAE AS6802 and the current TSN (Time Sensitive Networking) activities within IEEE.

Middleware Functions and Benefits

In this section we will give some more details on the middleware functionalities, usage and benefits.

- Communication: as noted above, our communication paradigm that has been implemented is AUTOSAR RTE based, with ultimately all internal signals mapped to Ethernet communication. The off-the-shelf Ethernet communication stacks provided by the OS suppliers are used, even though some specific enhancements have been made to incorporate the deterministic, time-triggered approach. The communication layer is generated at compile time by the SW integrator, based on signal information provided by the OEM and the SWC suppliers. All communication is made available to the outside world via the Deterministic Ethernet switch, thus giving total transparency for debugging and function performance analysis (for instance, inputs and outputs of the fusion layer can be made accessible without any timing impact on the system). As a benefit of the time-triggered approach, throughput and latencies are known in advance and do not vary depending on SWC behaviour. This also allows off-loading of SWCs to a PC via Ethernet (co-simulation) for prototyping or debugging purposes, while keeping all timing relations intact.

- Synchronization and time-triggered task scheduling: as for the communication, all major task activation is carried out in a deterministic way based on a predefined schedule. Violating SWCs which do not comply to their runtime restrictions are controlled and deferred. In some situations, this approach may seem too strict, therefore additional flexibility is built-in: SWCs may utilize other processing slots which are not fully used by the respective assigned SWCs, and SWCs may adjust their operation at runtime (dynamically) to the remaining time left in their scheduled slot. There are special API functions to enable both facilities, and they can mainly be used by functions which do not have a hard computational target, e.g. a fusion layer might adjust the number and granularity of objects identified depending on the available runtime.

- Safety: the middleware along with the underlying OS and HW mechanisms is conceptually implemented as a "Safety Element out of Context" (SEooC) in the sense of the ISO 26262, essentially providing a safe execution environment for each SWC and a safe communication channel to other SWCs. Generic safety mechanisms like memory protection, timing supervision and end-to-end communication protection between SWCs on same host and to remote hosts are derived from the established AUTOSAR mechanisms, and adapted and extended to the non-AUTOSAR environments by making use of the facilities provided by the respective operating systems and their host processors. Additional host-specific (HW depending) diagnostics

10

and supervision functions like clock, power supply, memory and uC core monitoring are implemented as necessary.

- Storage and data management: non-volatile and volatile memory is made available to the SWCs on each host in a location transparent manner, i.e. data written by an SWC on one host are transparently made available to all other SWCs on all hosts. Generic API functions as part of the middleware take care of storage, protection, data transport and retrieval across the ECU. This is again an essential feature to provide complete portability and location transparency of application SW components. In fact the SWCs are not even aware where data are actually stored – and do not need to care. This includes applications running on PC (SIL scenario).

- Diagnostics, Calibration, Flashing: these utility functions are implemented in a master/slave approach - the AUTOSAR based microcontroller acts as master, controls the other processing elements (slaves) and presents the whole ECU as one homogenous entity to the outside world. There is no need for a special tooling for diagnostics or calibration to cope with the internal diversity and complexity of the domain ECU, OEM and SWC suppliers may just keep their existing tools and workflow.

- SW development support: there is a full set of debugging and profiling tools available, which mostly have been derived from the specific OSs but enhanced by generic mechanisms (e.g. data logging). The PC based co-simulation environment cannot only be used as a convenient way to integrate prototyped SWCs under ADTF or MATLAB/Simulink with SWCs running on the target ECU. Identical C source code can be run on PC and on the target ECU without any timing-related impact, as the time-triggered mode of operation abstracts the usually much higher performance of the PC. Therefore, rich PC-based debug facilities may be used even for fully industrialized components, which would only offer very limited access to debugging and tracing facilities in a traditional workflow (in the embedded target).

Most of the middleware, in particular communication, is generated by a set of highly integrated tools and scripts, where the communication matrix and various additional information from OEM and SWC suppliers (in particular resource budgets, latency constraints and data flow constraints) are used as main inputs for a complete platform build. Tools from classic AUTOSAR BSW vendors and the OS suppliers are fully integrated to configure their respective components. Figure 4 gives an impression of the overall configuration and build workflow.

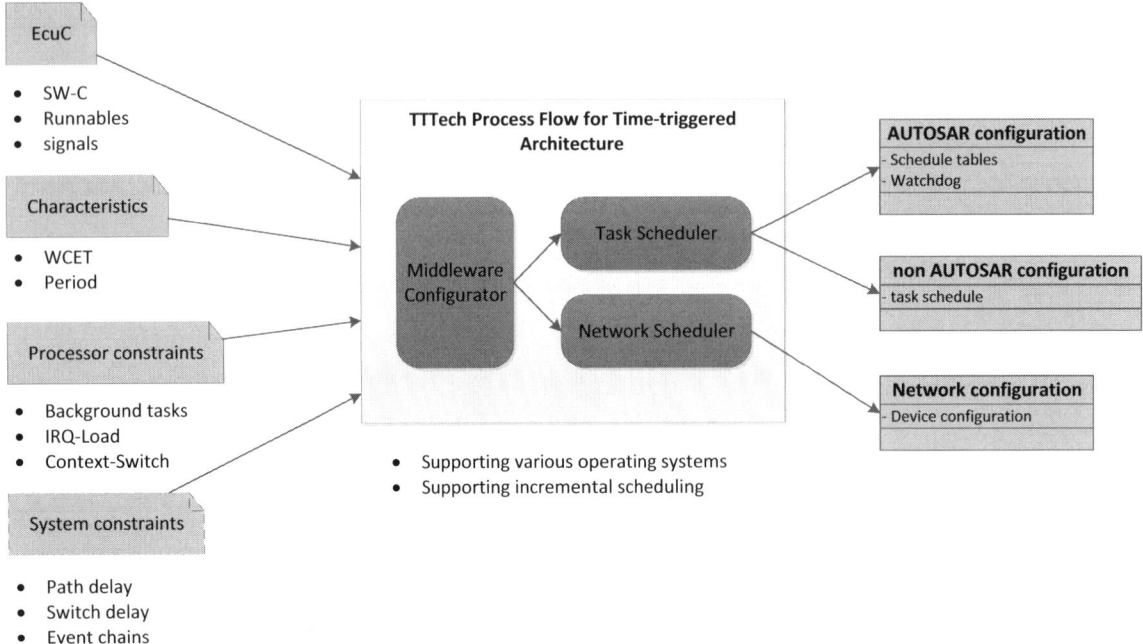

Fig. 4: Platform configuration workflow

SWC Development and Integration Workflow

In a traditional, ad hoc integration workflow all SWCs are delivered by the respective suppliers, a binary flash image is built for each host by the SW integrator and loaded into the system. Resource constraints (processing time, memory, communication bandwidth) are hopefully aligned before, but initially rarely met by the SWC suppliers; as a result, the SW integrator finds himself in a situation where total resource consumption is almost for sure out of limits and conflicts among SWCs arise. Overall functionality and stability will depend on the concrete inputs: even though simple environmental scenes might be treated smoothly, complex scenes will bring the system beyond its limits, leading to spurious faults and inconsistencies. Lots of mediation by the SW integrator and refinements by the SWC suppliers will be required to *iteratively* bring the system to a stable state. Worse yet, once the system is working, even minor changes by one of the SWC suppliers can destabilize the complete system again. This is a nightmare not just for the SW integrator, but also for the OEM's verification and validation tasks on a vehicle level.

Our architecture avoids this scenario all together and provides a stable, controlled process from individually tested SWCs to a completely integrated system. The deterministic, time-triggered scheduling of all communication and task activation provides the technical basis; it might seem rigid at first, but in essence just ensures the constraints that have to be met eventually anyway. So we are speaking of a conscious design step of strict resource planning and scheduling, to achieve overall system composability with the added benefit that all timing behaviour is known; even data flow latencies, that in a traditional workflow would have to be measured a posteriori or investigated by simulation and still would be prone to variability and jitter, are known precisely and in advance in our approach.

The actual SW integration workflow is carried out in 3 phases per SW release (Fig. 5), which may be triggered by OEM-defined SW integration steps and based on a functional feature release plan synchronized among the SWC suppliers.

1. Platform release: The SW integrator configures and builds the platform SW including all Basic SW, OSs and middleware and releases it to the SWC suppliers; resource boundaries (timing, memory) are predefined and aligned for each SWC and implemented through an appropriate task and communication schedule. All SWCs are "stubbed", i.e. replaced by empty templates and dummy functions.

2. (Single) SWC integration: the SWC supplier integrates his application on the platform by replacing the stubs and is able to test the SWC within its predefined boundaries, in exactly the same environment and timing as on the final integrated system; the tested SWCs along with the test cases used are delivered to the SW integrator for the system release.

3. System Release: SW integrator integrates all SWCs and delivers the complete build to the OEM and to the SWC suppliers for vehicle or test bench integration; all SWCs will run exactly as in the single-SWC setup (step 2), all test vectors are reused and results stay valid. The whole system is immediately stable, without any further measures.

(Actually there is also a step 0 which provides a so-called SIL release – an initial PC based environment for SWC development independent from the target.)

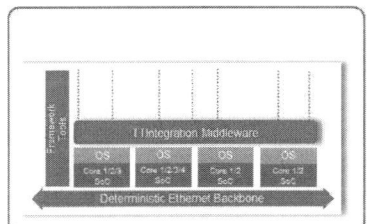 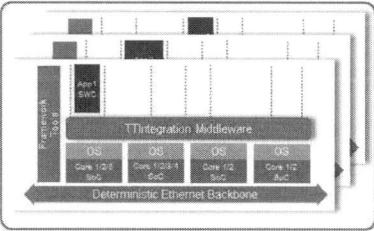 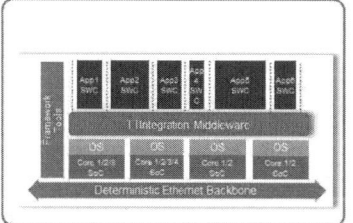

1. Platform Release.   2. (Single) SWC Integration   3. System Release

Fig. 5: SW Integration Process

13

At step 3, SWCs are checked if they fulfil their resource boundaries before integration ("acceptance test"), and possibly rejected if they do not[3]. The system may not be in a fully working state if other SWCs depend on the violator, but at least this is known in advance and corrective measures can be taken. In a traditional architecture and workflow, such violations might go unnoticed at first with simple scenes and low computational demand, and lead to an unstable system state later for example with complex environmental scenes – a situation much worse to deal with.

Each supplier can use the system release as a basis for further feature development and improvement, by incremental builds to upgrade his own SWC, as long as it obeys the defined boundaries.

To summarize, particular benefits and results of this controlled integration process (and the HW/SW architecture and middleware approach enabling it) are as follows:

- A running system without timing or memory collisions from SWCs is generated in one predictable, well-defined integration process.

- The actual runtime requirements of all SWCs are known and collected in an integration report.

- The actual data flow latencies and sequences of SWC activations are known and collected in an integration report (actually these "numbers" are known already after the 1st step – scheduling / platform release).

- Data flow latencies and all timing relations are not only known exactly, but are fixed – there is no variability or jitter, reducing the amount of testing and validation of the applications drastically.

- Side-effects from SWCs like blocking other SWCs are avoided.

Experiences from series development

The above mentioned HW architecture has been implemented in a real series project with very low impact on ECU cost: the only additional device needed besides the various processing engines is the Deterministic Ethernet switch, which has been integrated in an FPGA, whose processing capabilities and interface flexibility would have been needed anyway. An alternative is NXP's SJA1105T.

---

3 In an ideal world, this should not happen. However, to expedite SWC development in early project phases, strict timing checks might still be turned off because not all SWCs are available yet and spare time budget might be implicitly released to the existing SWCs.

The ECU is implemented in several variants to support different option packages from automatic emergency braking only up to autonomous driving functions. Due to the HW and OS abstracting properties of the middleware, SWCs are not aware on which variant they are running and may even be located on different hosts for different ECU variants. In addition, SWC suppliers do not have to spend any extra effort to support several ECU variants.

The SW Integration process is carried out smoothly by a surprisingly small team. The SWC suppliers have adapted quickly to the new process – likely because it is clearly defined and straightforward, gives SWC suppliers a well-defined development and integration platform with a multitude of useful features and relieves them from most of the housekeeping functions required in a traditional setup; they can simply focus on implementing the required algorithms, the required basis is delivered by the platform and middleware. The necessary precautions (estimating resource consumptions of SWCs and system-wide alignment of resource constraints in advance) are needed in any series development anyway and thus bring no additional effort.

The capability to move SWCs among different processing hosts without any code change is used quite frequently, especially since continuous refinement, learning and optimization over the SW integration steps reveal too optimistic initial resource estimates on some SWCs, which are offset by too pessimistic estimates on others – leading to changes in the allocation of SWCs over the course of the project.

Currently more than 30 applications from more than 10 suppliers are integrated – from sensory processing to various sensor fusion components to a multitude of customer functions like automated parking control and automated driving. It is our firm conviction that this project with its extreme technical and organizational complexity probably would not have succeeded without the implemented deterministic architecture, generic middleware and strict integration process.

Conclusion

A new architecture and integration approach for highly-integrated domain ECUs has been presented, which is currently being successfully implemented in the series development of a complex central ADAS controller containing a diverse set of processing units – from traditional automotive microcontrollers to high-end graphics engines. Two key elements enhance the architecture to fulfil all requirements on performance, safety, the development and integration process, and ECU modularity and scalability:

- A deterministic HW and SW architecture including a Deterministic Ethernet switch exploits the time-triggered paradigm for ECU internal communication and task scheduling, enabling predictable timing behaviour of application SW components and fully integrated systems.

- A common, generic AUTOSAR RTE-based middleware layer abstracts the diversity of the underlying microcontrollers and Operating Systems for the application SW components.

This architectural basis is implemented with an exhaustive feature set of debugging and utility functions and ensures full portability and location transparency for the application SW components, and therefore complete flexibility with respect to SWC allocation and system partitioning.

The integration process enabled by the deterministic, time-triggered paradigm guarantees full predictability and composability, and ensures smooth SW integration without iterative, time-consuming and costly integration hassle of traditional ad hoc approaches.

Optimized on human cognition – requirements and possibilities of digital exterior mirrors on commercial vehicles

Dipl.-Ing. Albert Zaindl

M. Sc. Linus Bundschuh

MAN Truck & Bus AG

© Springer Fachmedien Wiesbaden GmbH, ein Teil von Springer Nature 2018
R. Isermann (Hrsg.), *Fahrerassistenzsysteme 2016*, Proceedings,
https://doi.org/10.1007/978-3-658-21444-9_16

Introduction

Foreword: The first half of the paper summarises the author's contribution (Chapter 10) to the 'Handbook of Camera Monitor Systems – The Automotive Mirror-Replacement Technology based on ISO 16505', published by Springer in March 2016 [1]. This paper also uses some sentences from the authors' previous publications as well as from the authors' unpublished work, in some cases verbatim [1] [2] [3] [4]. For reasons of clarity the respective sentences are not cited separately.

Fig. 1 shows the field of view classes for a commercial vehicle. Six mirrors are currently required by law for commercial vehicles. These are the two main side view mirrors (Class II), two wide-angle side view mirrors (Class IV) as well as a close-proximity exterior mirror on the co-driver's side (Class V) and a front view mirror (Class VI).

Fig. 1: Field of view classes for commercial vehicles in accordance with ECE R46 [1]

An expected amendment to the regulation concerning devices for indirect vision, ECE R46 [5], in the course of this year, will permit camera monitor systems (CMS) to be used as replacements in all of these classes. The basis of the above is ISO 16505 as published in 2015 [6]. The amendment to ECE R46 specifies fundamental minimum requirements to be met by CMS. In each case, a comparison with the current mirror system is drawn. The main objective is that the CMS has to be on a par with the mirror.

In a joint research project, the Institute of Ergonomics (TUM) and MAN Truck & Bus AG (MAN) have developed a concept for replacing mirrors by camera monitor systems and have constructed prototypes. To achieve the most ergonomic concept possible, current utilisation of the mirrors by the driver was taken into consideration as well as the human capacity for visual perception. The result is a system that will considerably reduce the driver's stress and increase safety on the road, thanks solely to the way in which it depicts images.

Depicting the indirect fields of vision in a display gives rise to diverse possibilities for providing the driver with supplementary information. In particular, the augmented reality approach is a very promising one in CMS. This paper presents some approaches to driver assistance.

Human perception

To replace an existing system such as the current mirrors with a different technology in a sensible manner and even be able to generate added value in doing so, it is necessary to assess the utilisation of the current system and its properties in combination with human perception. In the specific case of mirror replacement, it is essential to grasp the fundamental concept of human visual perception.

Foveal and peripheral vision: the fundamental concept of visual perception

For a theoretical evaluation of human sight, two types of vision are of primary importance: foveal vision and peripheral vision [7]. The term 'foveal vision' describes the direct line of sight – in other words, precisely the area that is perceived consciously and with acuity. A high concentration of colour receptors (cones) in a very small area has an effect that is best compared with a telephoto lens. With a range of approximately 2 degrees, this area in which vision is sharpest forms a very small part (less than one thousandth) of the field of vision [8] [7]. Peripheral vision ('seeing out of the corner of one's eye') covers the greater part of the field of vision. It consists largely of rods and has very few cones by comparison. Moreover, the receptors here are not individually connected to an optic nerve as they are in the fovea, but instead, several are linked to a single nerve fibre. The characteristics of peripheral vision can be explained on the basis of this compression and the distribution of the photoreceptor cells. The primary activity is perception of contrasts with a high level of sensitivity to light and an efficient perception of movement.

Humans combine the advantages that arise from the interworking of both types of vision to form human perception. With the exception of scouting in the far distance, objects are as a rule detected with peripheral vision and then identified with foveal vision. In the course of this process, a pre-selection as to which objects we should focus on with foveal vision is already made subconsciously in the periphery. The decisive factors for the pre-selection include the intensity of the stimulus (the more intensive the movement/brightness of the object, the more likely it is for us to focus on it) as well as the expectation and experience of the respective person. An important factor in detecting objects in the peripheral field of vision is the so-called 'optical flow'. Whenever an object moves relative to the direction of the optical flow, it is detected. [8]

Table 1: Peripheral and foveal vision compared [1]

	Peripheral vision	Foveal vision
Percentage of visual field	>99.9%	<0.1%
Percentage of optic nerves	~50%	~50%
Type of sight	Brightness (black and white)	Colour vision
Photosensitivity	High photosensitivity	Low photosensitivity
Visual acuity	Low to very low visual acuity	High visual acuity
Cognitive level	Object detection	Object identification
	Good motion detection	Good object tracking (fixing)
	Random detection in a large visual field	Targeted 'scouting' in a very small visual field
	Parallel perception of objects	Fixed perception of one object
	Orientation in space	Exact targeting
Evolutionary purpose	Detection of objects/danger	Fixation onto objects/hunting

Perception in a mirror system

The peripheral detection of objects in a mirror system is made more difficult due to the different optical flows (described in more detail in [1]) in the main viewing direction and in the various mirrors. Instead of monitoring one global optical flow, the driver now has to keep track of seven different optical flows (the global basic flow plus the stipulated mirrors) simultaneously. As a consequent, perception in a mirror system entails the use of foveal vision (also shown in [9] [10]), which in turn means that drivers must deliberately direct their sight to every single mirror. This is not an issue for the primary rear-view mirror in a commercial vehicle because it is directed into the distance (to the rear) and thus corresponds perfectly to the evolutionary purpose of foveal vision, namely the targeted localisation and perception ('spotting') of distant objects (near the far point). However, the peripheral detection of objects in the wide-angle mirror is unlikely and even less likely in the close proximity mirror. In wide-angle and close proximity mirrors, the driver's task, namely the reliable, random detection of objects in the vicinity, which would ideally be accomplished with peripheral vision, must now be completed with foveal vision, i.e. with a conscious and time-consuming look at the respective mirror. [1]

4

Since it is only possible to focus in one viewing direction – foveal vision cannot work in a parallel fashion – the successive inspection of all individual mirrors requires a certain amount of time. Drivers need, for example, at least two seconds to check the three mirrors on the co-driver's side [11]. Meanwhile, at a speed of only 20 km/h the truck has already moved forward by eleven metres during which it has, for example, exceeded the length of the close proximity mirror's field of view more than twice over. As a result, experienced drivers have developed strategies that allow them to monitor what is happening on the road as effectively as possible under these difficult conditions. The strategies are specific to both drivers and situations, and are usually self-taught. However, even with these strategies in place it is often impossible for drivers to take in all the information around the vehicle in complex driving situations. Especially in complex traffic situations, drivers are forced to limit themselves to the field of vision they need for driving the vehicle (or, in other words, 'for accomplishing their objective'). It is up to the drivers to decide which mirrors they look at and which mirrors they leave unchecked. [1]

Combined fields of vision in digital external mirrors

In light of the above, it is clear that replacing each mirror with its own camera monitor system cannot be expedient. Instead of six mirrors there would be six individual images, which would leave the driver facing the same challenges as with the current mirror system. MAN, however, is applying an innovative display concept. In this concept, as Fig. 2 and Fig. 3 show, the viewing areas of the main side view mirror, wide-angle side view mirror and close proximity mirror (close proximity exterior mirror) on one side are combined into a single continuous display. Now the driver has to check only two monitors instead of five mirrors.

Fig. 2: CMS combines the three mirrors on the co-driver's side into a continuous display [3]

Fig. 3: In CMS, the cyclist is detected at first glance (via peripheral vision). In a mirror system, the driver has to shift his attention from the main exterior mirror to the wide-angle mirror in order to detect the cyclist.

The advantages of this view are obvious. The information that a driver would otherwise have to put together from three mirrors is presented in a single, comprehensive image. [12] This improves orientation and spatial localisation. The result is more efficient perception with less time and effort. Drivers no longer need to decide which of the three exterior mirrors they have to look at and can instead obtain all the necessary information at a single glance. Drivers can thus focus on their objective (scouting in the far distance or detecting the rear axle) using foveal vision while keeping all their close-range surroundings on the side in their peripheral vision. Objects can be detected much more quickly, more efficiently and, above all, even 'randomly'. The task of monitoring Areas IV and V, which ideally would be accomplished with peripheral vision, can now also be accomplished peripherally using the new system. [3] [1]

Besides the combined-display approach there are also approaches in which the fields of view are individually replaced. Fig. 4 shows various types of image display. The two pictures on the left correspond to a 1:1 replacement, the two pictures on the right are from the MAN approach. The example shows a classic turning-off situation. The systems are compared below on the basis of the images in the example.

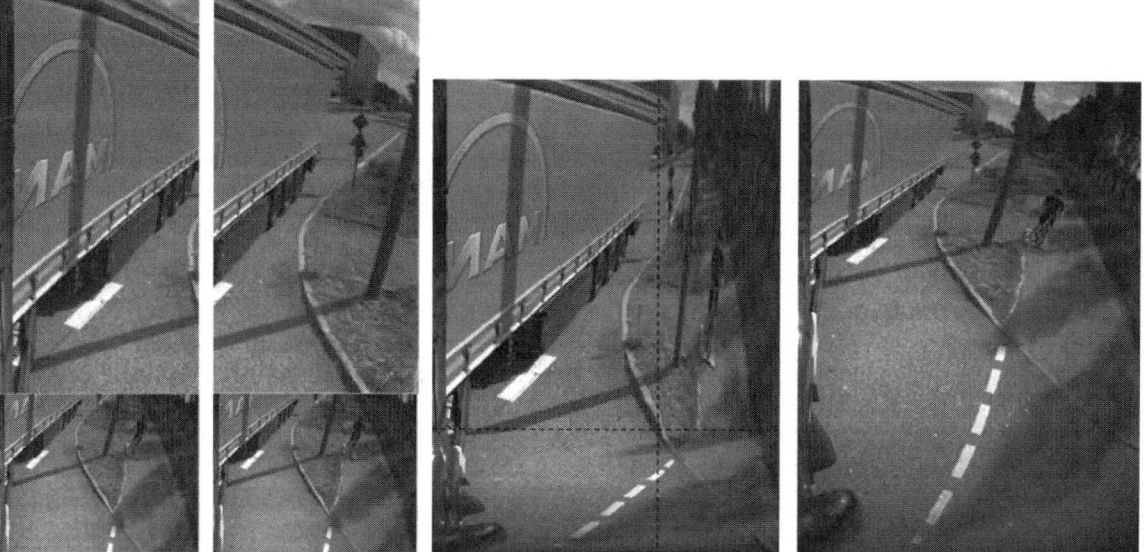

Fig. 4: Comparison of display options in mirror replacement systems (from left: 1:1 replacement, 1:1 replacement with panning function; combined replacement; combined replacement with manoeuvring function) [1]

In the conventional 1:1 replacement, the semitrailer hides the viewing area of the 'main mirror' (as is also the case in the current mirror system). Drivers are forced to shift their attention to the wide-angle image area. The cyclist can only be detected after this action is performed.

With a 1:1 replacement using camera tracking in the main mirror area, the rear axle of the semitrailer and the traffic behind it are clearly visible. The driver is no longer forced to look at the wide-angle area in order to determine whether the semitrailer will make it around the turn. This does, however, also pose a risk: if the driver fails to look at the wide-angle area, the cyclist will be overlooked completely.

In the third image in Fig. 4 (showing the combined view presented in this paper), the semitrailer again obscures the entire Class II field of view. However, the driver here can immediately detect the cyclist at a glance. Nonetheless, accurate manoeuvring around the turn is made difficult by the various distortions in the four areas.

In the manoeuvring view (right) the entire scene to the side of the vehicle is shown un-distorted. Thanks to the undistorted view, the driver can easily manoeuvre around the turn: The manoeuvring view displays the kink angle of the semitrailer and the overall situation in an ideal manner. The cyclist is detected right away and the cyclist's location in relation to the truck is immediately correctly assessed. The manoeuvring view is thus useful not only when manoeuvring: it can, for example, also assist the driver (if desired) in urban operation. [1]

Driver assistance and augmented reality

With the combination of viewing areas in the MAN display, drivers are able to perform monitoring tasks with a few glances that would previously have required re-focusing several times. One conceivable extension of this concept is obviously the provision of driver assistant system outputs and other information by the same means, so that driver information can be obtained as centrally as possible with minimal distraction from what is happening on the road.

Fig. 5: Warning in mirror-replacement monitor [4]

Fig. 5 shows a straightforward concept enabling the driver's attention to be drawn to the monitor. To raise the intensity of the warning, it can be pulsed or assisted by an acoustic warning. The colour selection takes aspects of human peripheral vision into account. [4]

However, the integration of augmented reality into the mirror-replacement display goes far beyond this. Augmented reality (AR) is the term used to describe a mixture of real and virtual environments that is closer to reality on Milgram's reality-virtuality (RV) continuum. [13]

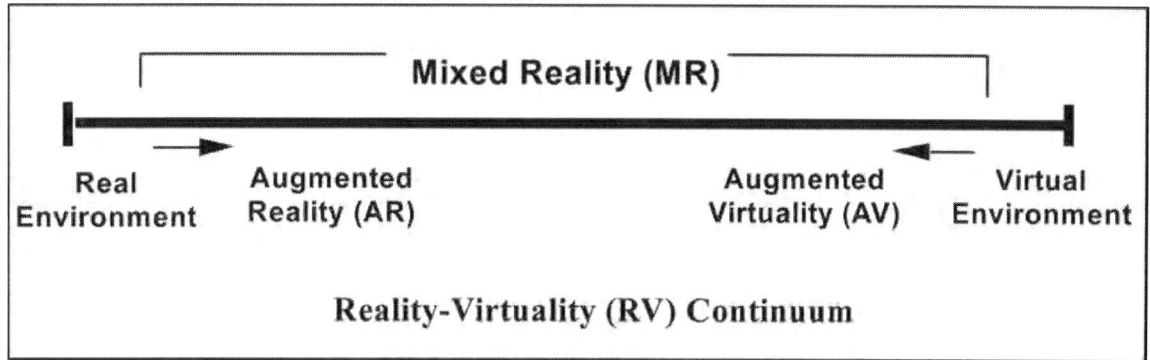

Fig. 6: Reality-virtuality continuum [13]

'Augmented reality is an extension of the perception of the real environment by means of virtual contents, which orientate themselves as far as possible to reality so that in extreme cases a distinction between real and virtual (sense) impressions is no longer possible.' [14]

In concrete terms, augmented reality in the context of mirror replacement can be described as the supplementation of the reality displayed by virtual contents in the correct perspectives. [4]

The objectives are primarily ergonomic. The expectation is therefore that information and warnings will be able to be grasped quickly and with spatial accuracy by the driver. Moreover, the driver's attention can be specifically controlled and behaviour with regard to mirror use improved. Finally, there is also an opportunity for developing adapted assistant systems and new types of assistant system on the basis of AR technology that simplify driving tasks and enhance safety. [4]

Some selected examples of concepts are presented below.

Display of vehicle's edges

Fig. 7: Display of line marking rear end of semitrailer to assist in changing lanes [4]

The current mirror system can sometimes lead to dangerous situations when a commercial vehicle changes lane on a motorway. One reason for this is that the speed of and distance from vehicles that are approaching from behind are often very difficult for drivers to estimate. Another reason is that due to the length of vehicle combinations, inadequate judgements as to whether a space in moving traffic is big enough for merging into are not uncommon. Displaying a line marking the length of the vehicle combination and stretching across to the neighbouring lanes can remedy this. This display enables the driver to assess the possibility of a lane change more easily. In this context it also makes sense if the marking line is activated only in certain driving situations, for example at speeds above 60 km/h or when a turn indicator is switched on. This would reduce the distraction of the driver by the overlay to the necessary minimum. [4]

Object detection and marking

Fig. 8: The cyclist is detected by the sensor system and marked in the CMS [4]

In certain driving situations and dangerous situations the driver can be assisted by the highlighting of objects in the vicinity. Under certain circumstances, additional data regarding the objects shown can be incorporated into the display. Decisive for a justified output of object information and markings is that the objects in question are actually relevant to the current situation. What this means in concrete terms is that, for example, pedestrians and cyclists who are at risk when the vehicle makes a turn, other vehicles in the case of a lane change, or obstacles where there is a risk of collision in constricted areas can be visually marked. When outputting data it is necessary to ensure it is as compact, clear and comprehensible as possible in every case. The decisive advantage of this type of data output is that, without much cognitive effort, the driver is immediately able to associate the information displayed with the correct object. This enables rapid reactions in dangerous situations. With an intelligent selection of colours (red, amber, green) it would also be possible to express the criticality of the situation. [4]

Display of virtual objects

Fig. 9: Using vehicle dynamics data, the image of a virtual semitrailer can be overlaid (e.g. for driving schools) [4]

Virtual objects that can be overlaid in the mirror-replacement system are other road users, obstacles or even a trailer/semitrailer. [...] Where a virtual trailer or semitrailer is used it must be taken into consideration that the handling of the virtual vehicle combination does not match that of a real combination. A better match can possibly be achieved by changes to the tractor vehicle's engine and brake management. In addition, with regard to operation in road traffic it must be noted that other road users are unable see the virtual trailer/semitrailer. For this reason they may not be able to categorize the mode of operation or may drive "through" the virtual trailer. Nevertheless, subject tests with manoeuvring tasks or driving-school exercises could be applications for the virtual trailer. [4]

Initial prototypical implementations

Fig. 10: Prototypical implementation of a manoeuvring and coupling aid [4]

For purposes of demonstration and to verify the technical feasibility, some of the concepts developed have already been implemented as prototypes by MAN (here in an early phase of the mirror-replacement prototype). The illustration shows a manoeuvring and coupling aid based on the steering angle, which assists drivers in accomplishing their objectives, similar to well-known series implementations of rear-view cameras. [4]

The integration of augmented reality in mirror-replacement systems is a highly complex task distinguished not only by technical but also and in particular by ergonomic requirements. In a survey of experts [4] AR concepts were seen predominantly as positive with regard to added value and enhanced safety. This shows that investment in the further development of this technology will pay off in the long run.

Summary and outlook

The replacement of conventional mirror systems by camera monitor systems will shortly be permitted by law. Especially for commercial vehicles, an enormous potential – particularly with regard to safety – arises from the optimisation of human perception. As described above, MAN's mirror-replacement system is optimised for human perception. The intuitive layout of the image in CMS makes it easier for the driver to analyse objects in relation to the vehicle and to follow these objects beyond the boundaries of the respective fields of vision. Drivers are offered a single comprehensive image featuring all the information that they would otherwise have to put together from three individual mirrors on each side.

The display in a camera monitor system gives rise to additional possibilities for providing drivers with information about what is happening on the road. One promising approach here is augmented reality, i.e. the extended representation within the real event. The driver thus receives the information exactly where it is needed. The expectation is therefore that information and warnings will be able to be grasped quickly and with spatial accuracy by the driver. Moreover, the driver's attention can be specifically controlled and behaviour with regard to mirror use improved. In combination with other sensor systems, it may in future even be possible to display objects in CMS that the driver is not yet able to see. Finally, there is also an opportunity for developing adapted assistant systems and new types of assistant system on the basis of AR technology that simplify driving tasks and enhance safety.

Bibliography

[1] A. Zaindl, "Camera Monitor Systems Optimized on Human Cognition— Fundamentals of Optical Perception and Requirements for Mirror Replacements in Commercial Vehicles," in *Handbook of Camera Monitor Systems – The Automotive Mirror-Replacement Technology based on ISO 16505*, Berlin, Springer , 2016, pp. 313 – 328.

[2] A. Zaindl, "Spiegelersatz am Nutzfahrzeug," *Ergonomie Aktuell,* pp. 30-33, 2015.

[3] A. Zaindl, A. Zimmermann, K. Dörner and C. Kohrs, "Camera-Monitor System as Mirror Replacement in Commercial Vehicles," *ATZ*, pp. 12-17, 05 2015.

[4] L. Bundschuh, *Erweiterter Spiegelersatz: Konzeptentwicklung für den Einsatz von Augmented Reality,* MAN Truck & Bus AG and TU München, Lehrstuhl für Ergonomie, Eds., München, 2014.

[5] UN/ECE, *Regelung Nr. 46 der Wirtschaftskommission der Vereinten Nationen für Europa (UN/ECE) — Einheitliche Bedingungen für die Genehmigung von Einrichtungen für indirekte Sicht und von Kraftfahrzeugen hinsichtlich der Anbringung solcher Einrichtungen,* 2009.

[6] ISO 16505:2015, "Road vehicles -- Ergonomic and performance aspects of Camera Monitor Systems -- Requirements and test procedures," International Organization for Standardization, 2015.

[7] H.-W. Hunziker, Im Auge des Lesers: foveale und periphere Wahrnehmung – vom Buchstabieren zur Lesefreude, Zürich: Transmedia Stäubli Verlag, 2006.

[8] V. Gralla, Peripheres Sehen im Sport – Möglichkeiten und Grenzen dargestellt am Beispiel der synchronoptischen Wahrnehmung, Bochum: Ruhr-Universität Bochum, 2007.

[9] R. Rassl, Visuelle Informationsaufnahme des motorisierten Verkehrsteilnehmers – Visuelle Informationsaufnahme des motorisierten Verkehrsteilnehmers, München: Technische Universität München, 2004.

[10] J. Hudelmaier, Sichtanalyse im Pkw – unter Berücksichtigung von Bewegung und individuellen Körpercharakteristika, München: Technische Universität München, 2003.

[11] A. Zaindl, A. Zimmermann and K. Bengler, "Untersuchung der Spiegelnutzung im Nutzfahrzeug," in *Nutzfahrzeugworkshop* , Graz, 2014.

[12] K. Bengler, M. Götze, L. Pfannmüller and A. Zaindl, "To See or not to See-Innovative Display Technologies as Enablers for Ergonomic Cockpit Concepts," in *Electronic Displays Conference*, Nürnberg, 2015.

[13] P. Milgram, H. Takemura, A. Utsumi and F. Kishino, *Augmented Reality: A class of displays on the reality-virtuality continuum,* Toronto, Canada, 1994.

[14] R. Dörner, W. Broll, P. Grimm and B. Jung, Virtual and Augmented Reality (VR/AR), Berlin Heidelberg: Springer, 2013.

Impact of driver assistance systems on future E/E architectures for commercial vehicles

Dr. Andreas Lapp, Thomas Thiel, Michael Schaffert

Dr.-Ing. Andreas Lapp, **Chief Architect**, Systems Engineering for Commercial Vehicles & Off-Road Applications, Robert Bosch GmbH, Postfach 30 02 20, 70442 Stuttgart, Tel. +49(711)811-44405, Andreas.Lapp@de.bosch.com

Dipl.-Ing. (FH) Thomas Thiel, Director Engineering, Systems Engineering for Commercial Vehicles & Off-Road Applications, Robert Bosch GmbH, Postfach 30 02 20, 70442 Stuttgart, Tel. +49(711)811-22476, Thomas.Thiel@de.bosch.com

Dipl.-Ing. Michael Schaffert, Vice President, Leader Center of Competence E/E-Architecture, Robert Bosch GmbH, Robert-Bosch-Allee 1, 74323 Abstatt, Tel. +49(7062)911-6057, Michael.Schaffert@de.bosch.com

© Springer Fachmedien Wiesbaden GmbH, ein Teil von Springer Nature 2018
R. Isermann (Hrsg.), *Fahrerassistenzsysteme 2016*, Proceedings,
https://doi.org/10.1007/978-3-658-21444-9_17

1 Abstract

New driver assistance systems on their way to automated driving are pushing the currently available E/E architectures to the limits. On the one hand, this applies to the domain and ECU topology, which will be extended by additional components, e.g., for environmental sensing, data processing, automated vehicle control and for the required surround view. On the other hand, redundant communication and system functions are necessary to ensure a "fail-degraded" or "fail-operational" system behavior. These redundancy requirements also apply to the vehicle power supply so that a redundant power supply is available to ensure that the braking system and steering system remain available even in the event of failure.

This paper introduces an E/E architecture proposal for Commercial Vehicles (medium & heavy duty) which supports the automated driving functions such as Traffic Jam Pilot, Highway Pilot and Pairing/Platooning[1] and discusses the necessary redundancy concepts for communication, system functions and power supply.

Technological enablers for economic implementation of the necessary redundancies, as well as zone-based structures based on central communication backbones and energy backbones, are also proposed that offer potential for simplifying wiring harnesses available today.

Since connecting the vehicle to the Internet and to other vehicles or to the vehicle environment is also part of automated driving systems, such as Pairing for instance, a multi-level security concept for automotive applications is also presented that secures the vehicle against external attacks and provides multi-level security to the functions and communication in the vehicle. The resulting structural requirements on E/E architectures are also represented.

2 Future applications as challenge for the E/E architecture

The key driver for future E/E architectures in MD/HD commercial vehicles are applications that significantly extend the functionality in today's vehicles. Examples are shown in Figure 1, "Functionality":

- **Automated Driving functions** such as Traffic Jam Pilot, Highway Pilot and Platooning require automated longitudinal and lateral control. For this, a reliable sensor

---

1 Platooning = Sequence of multiple vehicles, Pairing = Platooning /w two vehicles

2

system is required that detects the environment at the front and sides of the vehicle, and in case of an automated lane change, even the environment at the rear of the vehicle. However, lack of integration of the trailer in the E/E architecture of the truck is hampering 360° surround view. Although the legal regulations for Automated Driving are still under discussion, at least a "fail-degraded" approach is needed. In other words, depending on degree of automation, the driver or an electronic system must transfer the vehicle to a safe state.

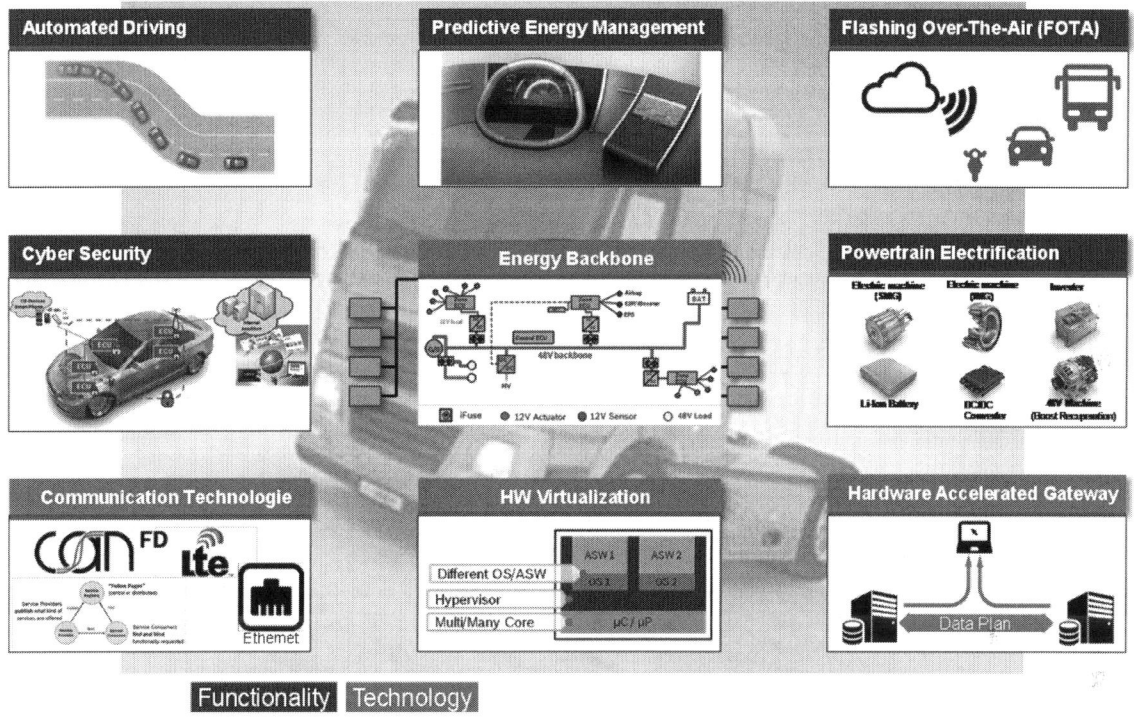

Figure 1: Innovative functionality and technologies that affect the E/E architecture

- **Predictive Energy Management:** cross-domain, holistic approach focusing on reducing the fuel consumption. This requires sensor data together with self-learning map data; a fast data connection to the outside is a basic condition here.
- **Flashing-Over-The-Air (FOTA):** wireless update capability for all ECUs outside the workshop taking into account any safety requirements while maintaining the procedures established by the OEM.
- **Cyber Security:** multistage, holistic approach to prevent both external attacks on the vehicle and to ensure the confidentiality and integrity of all communications.
- **Powertrain Electrification:** depending on the degree of electrification, vehicle electrification enables reduction in fuel consumption through energy regeneration and boosting and/or a reduction in exhaust gas emissions and noise.

Beyond the functionality shown in Figure 1, there are also further innovations, such as **preventive diagnosis**. This monitors the "health" of the main vehicle components and computes a remaining service life prediction so that components can be replaced at the right time and at a logistically appropriate time. The major goal is to ensure the availability of the vehicle.

Figure 2 shows the development of E/E architectures, starting from distributed architecture, followed by domain-centralized architecture, and finally the vehicle-centralized or zone-based architecture. Currently we are, depending on the OEM, at the transition zone between the distributed architectures and domain-centralized architectures.

While the development evolves from a modular, distributed architecture to a centralized architecture within the established functional vehicle domains, the step towards a zone architecture represents a revolutionary step. Zone-oriented architectures have the advantage of significantly reducing the complexity of the wiring harnesses and the cable lengths associated with them. Consequently, functions no longer need to be assigned mainly based on their relevance to a functional domain, but can be assigned based on the installation location in the vehicle. Moreover, centralization of vehicle functions and domain functions facilitates vehicle-wide optimization, e.g., in the scope of energy management system.

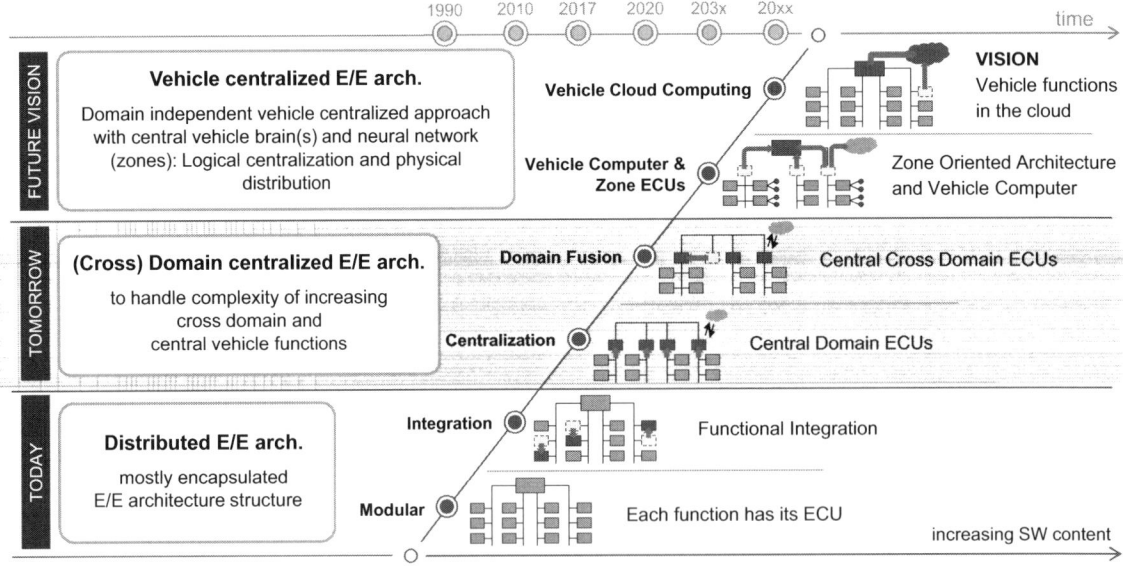

Figure 2: (R)Evolution of the E/E architecture

There will be a rise in the connectivity to the "Cloud" and transfer of functions from the vehicle to the cloud, irrespective of the architecture paradigm used.

So far the development was driven by the passenger car segment; contrary to the passenger car segment, the introduction of driver assistance systems in commercial vehicles is driven by the profitability for the driver or fleet operator. From this point of view, it is expected that the commercial vehicle will be the "front runner" for the introduction of individual functions and business models.

The hitherto evolutionary development and projection presented here can also be revolutionized, at least to some extent, with the availability of disruptive technologies or approaches.

3 Today's architectures

Figure 3 shows two typical E/E architecture paradigms as seen in the MD/HD commercial vehicles available in series production today. On the one hand, "star topologies" are available in which domain specific subnetworks branch from a central vehicle control unit. This vehicle control unit can host vehicle and domain functions (e.g., Energy Management, Powertrain Management, Environment Modeling and strategy for Automated Driving) and a central gateway.

On the other hand, "backbone architectures" are available in which the domain control units decouple the domain-specific sub-networks from the rest of the vehicle. These domain control units are connected via a backbone bus (typically CAN) and respectively contain the necessary gateway functionality. The two architecture paradigms are mainly characterized by the following aspects:

- Domain-based design
- Vehicle control units or domain control units control single or multiple domains, cross-domain functions are usually hardly available.
- Simple assistance systems that require few sensors and limited functionality.
- Redundancies are present where they are required by law
- Cyber-security concepts are partially available locally at the interfaces to the vehicle environment
- Trailer integration at low technological and functional level

Generally, the vehicle OEM is responsible for the E/E architecture. It follows the OEM-specific innovation cycles, the OEM also acts as a system integrator.

In particular, multitude of variants are reflected as large expenditure for systematization and standardization and for variant-specific adaptation. In addition, different markets are equipped with different architecture variants due to other applications, pricing policies etc.

Last but not least, to an extent the E/E architecture also mirrors the organizational setup of the respective OEM.

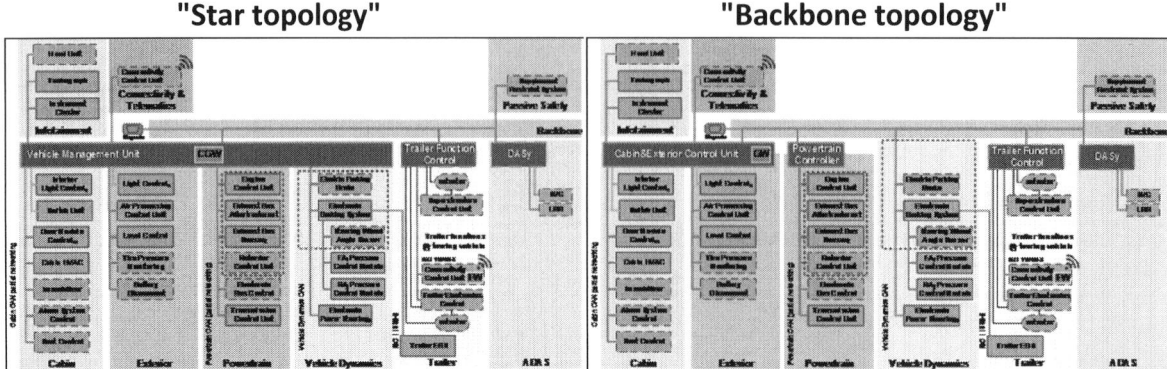

Figure 3: Examples for today's architecture topologies

4 Innovative technologies

In this section, exemplary innovative technologies are presented that are necessary for introducing or supporting the functions presented in section 2 (see also Figure 1, "Technology").

- **Energy backbone:** When introducing zone oriented architectures, it is advisable to connect the individual zones in the vehicle via an energy backbone to reduce the complexity of the wiring harness and the necessary cable lengths [10]. Cable lengths can be further significantly reduced by the introduction of intelligent semiconductor switches and fuses, because then one or more central fuse boxes, accessible by the driver, can be omitted. Semiconductor components also increase the reliability and availability, moreover, they provide additional functionality (e.g., current measurement, free choice of fuse characteristic). The additional costs required for semiconductor components must be seen in the scope of a TCO analysis. The backbone can be raised to a higher voltage level (for example 48V) in order to reduce the losses or to reduce the cross-section of the cables.

- **Communication Technologies:** The rising demands on the intra-vehicle and inter-vehicle communication are leading to the introduction of CAN FD [4], Ethernet [7] for the internal and LTE (Long Term Evolution [1]) for external vehicle communication. It would be advantageous to introduce a service-oriented communication (SoC) in the vehicle to enhance the general communication flexibility, for instance between truck and trailer . First automotive specific SoC-implementation is defined in AUTOSAR [2].

- **Hardware Virtualization:** With hardware virtualization, independent blocks for handling software variance are created within one control unit. The software can be encapsulated inside these blocks for efficient software reuse and software sharing. It can also be used to run the software with different safety or time requirements and legacy software, efficiently on a control unit while keeping the validation effort manageable by a delta concept [8]. Variants useable on μP or μC are available from Bosch.

- **Hardware Accelerated Gateway:** Hardware accelerated gateways help maintain a separation between data and control flow when data is transferred from one bus to another. The control unit CPU then needs to address the control flow only and hence will be relieved of load significantly. At the same time, the data flow is implemented by a dedicated hardware with high throughput, very low latency and negligible jitter. Overall, this leads to an efficient and high-performance data processing while minimizing CPU load [8].

5 Architectural Drivers

The implementation of the new functionalities presented in section 2 and the introduction of new technologies (section 4) require adjustments of today's E/E architectures. But implementation of new functional requirements is not the only factor involved here. In particular, future architectures are characterized by qualitative requirements. So an exponential increase in internal communication is to be expected, and the external communication will increase as well significantly due to the introduction of corresponding technologies, such as LTE. Similarly, demands on computing power and the software complexity per control unit will increase, however, without increasing the number of control units.

Additional requirements such as functional safety, security, availability, variance, consideration of take-rates, essentially determine the required control units, the topology and the selection of bus systems.

Consideration of qualitative requirements is characterized by trade-offs:

- Unit costs versus development costs
- Innovation versus legacy
- Evolution versus revolution
- "balcony building" versus structural changes

As part of a systematic, structured approach, a functional network of interactions will be developed based on use cases and the resulting requirements [9]. This functional network serves as a base for deriving an E/E architecture taking into account the innovative technologies (see section 4) and the qualitative requirements described earlier.

6 Future E/E architectures

The E/E architecture proposal illustrated in Figure 4 is based on the premises listed below:

- Support to the functions introduced in section 2, such as Automated Driving, Electrification, Cyber Security, Predictive Energy Management, Predictive Diagnostics, FOTA
- Scalability of E/E architecture for different levels of Automated Driving and Electrification
- "Fail-degraded" functionality for Automated Driving
- Holistic networking of the trailer to the truck E/E architecture

Figure 4 shows a backbone architecture in which the domain ECUs are connected with each other by Ethernet to meet the growing communication needs, such as those required for FOTA, Predictive Diagnosis and Automated Driving. It represents an evolutionary development of modern architectures, in which the proven functional domains remain available.

Some selected aspects are explained below with examples.

Powertrain

Using a domain controler for the "Powertrain" domain offers several advantages. Firstly, it can accommodate the vehicle functionality and thus enables the consequent logical separation of engine functionality that is hosted on the Engine Control Unit. This facilitates business models, in which the internal combustion engine including associated ECUs and sensors can be purchased or sold as single unit.

Secondly, it allows the decoupling of the "Powertrain" domain from the rest of the vehicle and thus allows the flexible addition of ECUs, for example, to the powertrain electrification or to improve energy efficiency via waste heat recovery. However, those benefits can also be achieved by using a star topology with a vehicle control unit. But if a predictive energy management with operating point optimization or a predictive diagnostics with more elaborate pre-processing are to be implemented in the vehicle, the necessary processing power requires a separate domain ECU for the powertrain.

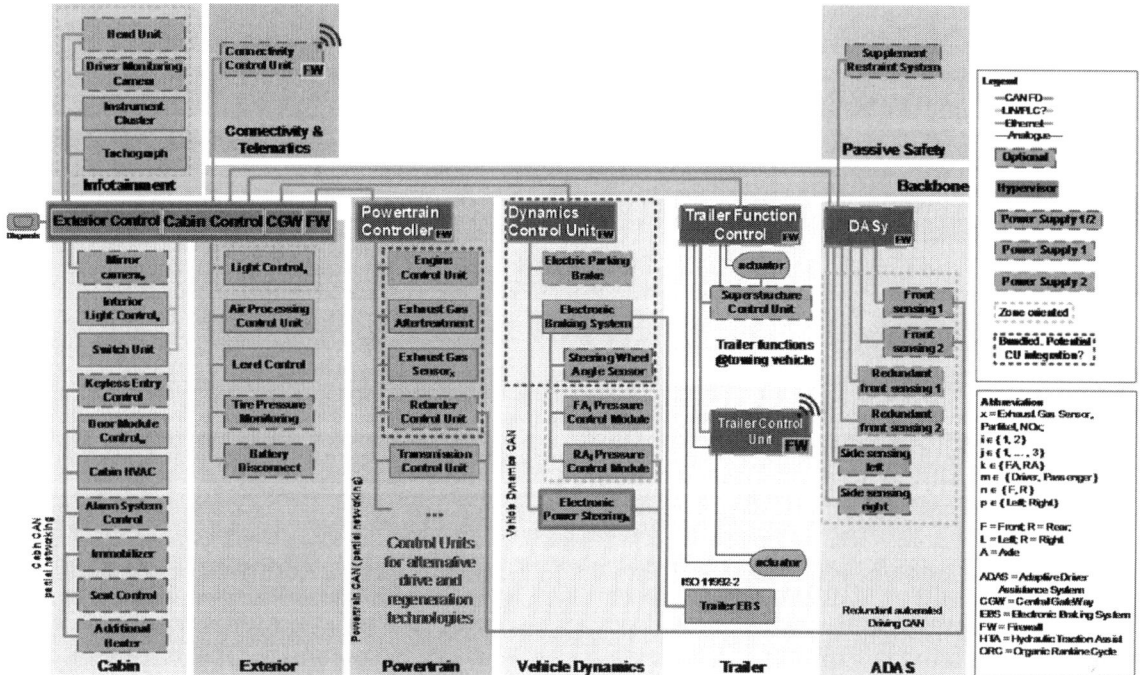

Figure 4: E/E architecture proposal for commercial vehicles (MD, HD) 202x for automated driving and powertrain electrification: Backbone topology.

Connectivity

Connecting the vehicle by means of a "Connectivity Control Unit" (as a stand-alone device or integrated in a Central Gateway for instance) via an air interface is required for Flashing-Over-The-Air, but also for the implementation of parts of a predictive energy management and for predictive diagnostics.

Cyber-Security

The topic "Cyber-Security", which is not only required for safeguarding the vehicle connectivity but as well for the diagnostic interface, is in Figure 4 just indicated by the blocks with the abbreviation "FW" (firewall). The description of the 4-level security concept of Bosch is provided in [8].

Automated Driving

The automated driving is implemented by adding the ADAS-domains (Adaptive Driver Assistance System). Here, it is possible to scale starting with an assistance system, via semi-automated and highly automated, finally to fully automated systems. The sensors required are decoupled from the remaining vehicle via the domain ECU "DASy" (Driver Assistance System). The DASy provides the necessary computing power for accom-

9

modating the environment modeling and the automated driving control, consisting of trajectories management and a coordinated actuator management. Although the legal regulations for automated driving are still under discussion, it can be expected that for higher levels of automation the system must be able to stop the vehicle at a defined safe place (e.g. emergency lanes) and secure it against rolling [6]. This is also referred to as "fail-degraded".

For such a fail-degraded system behavior, the design should not only include redundancy for sensors, actuators and the control functionality, but also for the communication between the ECUs involved and the vehicle power supply [3, 5]. This is taken into account in the architecture, shown in Figure 4:

- Redundant sensors for environment detection based on diverse technologies within the ADAS domain (combination of video, lidar, radar).
- Redundant actuators for acceleration, deceleration and steering in the "Powertrain" and "Vehicle Dynamics" domains.
- Redundant control functionality in sensor ECUs (Front Sensing 1, 2).
- Redundant, diverse communication (CAN) between control system and actuators.
- Relevant components are safeguarded by a second power supply unit (Power Supply 1 and 2).

Trailer

The trailer ECUs on the truck (Trailer Function Control) and on the trailer (Trailer Control Unit) are upgraded as integration platforms, so that further functions for superstructures, attachments and trailers (e.g. locating the trailer, operating cooling unit, tire pressure control, coupling to industry 4.0 applications) can be accommodated. To this end, the Bosch trailer ECUs are equipped with extensive transmission and reception technologies such as GNSS (Global Navigation Satellite System), Cellular Network, WLAN, Bluetooth®, pWLAN, 433MHz receiver.

For coupling between the trailer and truck, a CAN connection is provided. Coupling via WLAN is also possible but problematic because the reliability and availability must be ensured. Embedded Service Oriented Communication (eSoC) (see section 4) is suitable for flexible integration of the functions of superstructures, attachments and trailers. Furthermore, additional sensors are required on the trailer for fully automated driving, for example, radar sensors or cameras. The information gained here must be provided in real time to the truck which requires a highly performant (e.g. Ethernet), reliable, integration of the trailer.

Star versus backbone topology

The architecture shown in Figure 4 can be implemented in a star topology with comparable performance features (see Figure 3, figure to the left) by using a vehicle control unit. In such a case, the "Trailer" and "ADAS" domains should continue to be represented as separate domain ECUs due to reasons of flexibility and resource requirements.

The advantage offered by a backbone architecture is that hardware variants, required due to different functional content, take-rate or innovation cycles can easily be implemented. Hence, additional functions can be added or, depending on the resource requirements, domain ECUs with scalable computing power can be used. Moreover, a backbone architecture with multiple domain ECUs offers advantages in the implementation of redundancies required.

A star topology with a vehicle control unit is less flexible regarding the hardware variance, but usually offers cost advantage when the functionality is fixed. Here also software-variance can be mapped by hardware virtualization.

7 Summary

In this paper, a future E/E architecture is proposed that supports innovative functionality such as Automated Driving, Electrification, Cyber Security, Predictive Energy Management, Predictive Diagnostics, FOTA.

The architecture is derived evolutionarily, using a systematic approach, from today's architectures and covers both redundancy concepts for Automated Driving and other requirements, such as scalability.

Furthermore, innovative technologies are highlighted that easily facilitate the introduction of new functionality and which are required or help reduce the development efforts.

8 References

[1]   3rd Generation Partnership Project (3GPP), LTE, http://www.3gpp.org/

[2]   AUTOSAR Partnership, Spezifikationen 4.x,
      http://www.autosar.org/specifications/

[3]   Blume, Fausten, *Unfallfreies Fahren durch Automatisierung*, 15. VDA Kongresss
      Autonomes Fahren, 03.2013

[4]   Bosch, *CAN with Flexible Data-Rate*, Specification, Version 1.0, 04.2012

[5]   Fausten, *Challenges on the Way to Automated Driving*, 16. Zulieferertag Automo-
      bil Baden-Württemberg, 11.2014

[6]   Gasser, u.w., Gemeinsamer Schlussbericht: BASt-Projektgruppe „Rechtsfolgen
      zunehmender Fahrzeugautomatisierung", Berichte der Bundesanstalt für Straßen-
      wesen Fahrzeugtechnik, Heft F 83, 01.2012

[7]   IEEE802.3 100BASE-T1, http://www.ieee802.org/3/index.html

[8]   Klauda, Schaffert, Lagospiris, Piel, Kappel, Ihle, *Weichenstellung für 2020 –
      Paradigmenwechsel in der E/E-Architektur*, ATZ Elektronik, 02.2015

[9]   Lapp, Thiel, Lagospiris, *Use-Case basiertes Vorgehen zur Ableitung zukünftiger
      E/E Architekturen am Beispiel Nutzfahrzeug*, 4. Internationales Commercial Vehi-
      cle Technology Symposium, Kaiserslautern, 03.2019

[10] Pieraccini, Pflüger, *Powering the future*, 17. International Congress ELECTRON-
      ICS IN VEHICLES, Baden-Baden, 10.2015

Tagungsbericht

Markus Schöttle

© Springer Fachmedien Wiesbaden GmbH, ein Teil von Springer Nature 2018
R. Isermann (Hrsg.), *Fahrerassistenzsysteme 2016*, Proceedings,
https://doi.org/10.1007/978-3-658-21444-9_18

2. ATZ-Tagung Fahrerassistenz
Iterative Schritte ohne Zeitzusagen

Auf der 2. Internationalen ATZ- Fachtagung Fahrerassistenzsysteme trafen sich am 13. und 14. April über 200 Experten in Frankfurt am Main. Neben der Bandbreite aller relevanten Themen wie Märkte und Nutzer, Sensorik, Komponenten und E/E-Architekturen legte die von Etas und Continental unterstützte Tagung dieses Jahr einen Schwerpunkt auf die noch vergleichsweise wenig ausgeprägte Disziplin Automotive IT-Security im und um das vernetzte Fahrzeug.

EINIGE FRAGEN SIND NOCH UNGELÖST

Im Gesellschaftshaus Palmengarten in Frankfurt am Main gab Elmar Frickenstein im Rahmen seiner Keynote zur 2. Internationalen ATZ-Fahrerassistenztagung eine ehrliche und offene Einschätzung der Herausforderungen auf dem Weg zum automatisierten Fahren. Als langjähriger Leiter Elektrik/Elektronik und jüngst ernannter Bereichsleiter Vollautomatisiertes Fahren und Fahrerassistenz der BMW Group weiß Frickenstein mit zu hohen Erwartungen umzugehen: „Vor vier Jahren bin ich mit einem hochautomatisierten Fahrzeug von München nach Ingolstadt gefahren", sagte er und ließ keinen Zweifel: „Das wird kommen. Wir werden den Weg unbeirrt gehen, dennoch gilt es, sich in iterativen Schritten an die technischen Realisierungen für den Serieneinsatz heranzutasten". Demnach bleibe es in einigen technischen Disziplinen noch offen, wann sich diese realisieren lassen. „Einige wichtige Fragen sind noch ungelöst", sagte Frickenstein

DIE DATENFLUT WIRD ZUNEHMEN

„Die Automobilhersteller sind in Anbetracht der offenen Fragen abhängig von den Fortschritten und der erfolgreichen Zusammenarbeit mit anderen Branchen der IT- und Consumer-Elektronik-Industrie", betont Frickenstein. Machine Learning und Supercomputer seien für das sichere automatisierte Fahren notwendig. Mit jedem vernetzten Fahrzeug mehr komme in diesem Backend eine gewaltige Datenlast zusammen. Das BMW-Backend verarbeitet beispielsweise aktuell 200 Pings pro Sekunde nur für den Dienst RTTI (Real Time Traffic Information). „Die vernetzte BMW-Flotte ist heute in der Lage 95 % aller Autobahnen abzufahren", berichtete Frickenstein. So sei das Auto Teil der Cloud und liefere ständig Daten. Doch die Datenflut wird noch zunehmen, was mit heutiger IT-Performance nicht mehr darstellbar ist. Das Testen der Systeme im Real Traffic wird nach Meinung des BMW-Manns die IT-Zukunftstechnologie weit vorher benötigen. Zu einem der Unternehmen, die dies ermöglichen müssen, zählt IBM. Erich Nickel, Geschäftsführer des IBM Automotive Center of Competence weiß um die Herausforderungen und schätzt diese allerdings als lösbar ein. Sowohl Rechnerkapazitäten als auch

Rechengeschwindigkeiten würden in den kommenden Jahren so erhöht, dass sich die Ziele der Fahrzeugvernetzung und Hochautomatisierung realisieren ließen.

SECURITY – EIN SENSIBLES THEMA

Insbesondere die Rechnergeschwindigkeit ist eine unschätzbare Ressource, wenn es um Safety und Security geht, zumal das schnellstmögliche Ausschalten von Hackerangriffen gewährleistet sein muss. Nickel erklärte dies in seinem Vortrag „Der Kampf gegen Cyber-Attacken auf vernetzte Fahrzeuge". Security ist ein sensibles Thema, ein großes Lernfeld selbst für die Spezialisten, die bereits in anderen Branchen, beispielsweise Banken und Versicherungen seit Jahrzehnten täglich damit zu tun haben und sich nun auf die Anforderungen, im automobilen Umfeld einrichten müssen. Zu diesen zählt die Firma Utimaco. Der deutsche Security-Spezialist sichert unter anderem die bidirektionale Vernetzung zwischen Kundenfahrzeug und dem amerikanischen Elektroautohersteller Tesla unter strengen Security-Maßnahmen ab. Geschäftsführer Malte Pollmann erklärte in der Podiumsdiskussion: „Wir liefern Verschlüsselungsgeräte für sensitive, schützenswerte Daten, sogenannte Hardware-Security-Module, HSM, mittlerweile auch an viele Automobilhersteller". Dennoch betont er, dass er die Sicherheitskonzepte und notwendigen Architekturen, bei Tesla organisatorisch und prozessual seit Jahren verankert, als vorbildlich erachtet.

VORBILD USA

Zumindest beeindruckend ist die Positionierung und der Pragmatismus des US-amerikanischen Herstellers, der in Anbetracht kleinerer Stückzahlen weniger Risiko eingeht und den klassischen Automobilbau nicht bedienen muss. Dennoch mache Tesla Fehler. In der Diskussion auf dem Podium wie auch in Vorträgen wurde die durchaus kritisch zu beurteilende Vernetzung von Tesla-Fahrzeugen nicht ange-sprochen. Doch in Hintergrundgesprächen sprach man über lebensgefährliche Fahr-manöver, die man selbst mit Testfahrzeugen der Marke erlebt habe, ausgelöst durch fehlerhafte Fahrerassistenzsysteme. Wenn diese auch noch durch Updates nachträglich „over the air" geordert und aktiviert werden, müsste nach Meinung einiger Experten eigentlich die Fahrerlaubnis erlöschen. Pragmatismus macht US-amerikanische Unternehmen und die Regionen, in denen sie forschen und entwickeln, zu einem interessanten und willkommenen Innovationsmotor in Zeiten der automobilen Digitalisierung, Elektrifizierung und Automation. So folgte Dr. André Weimerskirch der Einladung von ATZlive, um aus den USA zu berichten. Der deutsche Wissenschaftler leitet als Security-Spezialist der Universität Michigan einen Feldversuch mit hochautomatisierten Fahrzeugen im Auftrag der amerikanischen Verkehrssicherheitsbehörde. Er kennt sowohl das international anerkannte Netzwerken der deutschen Automobilindustrie als auch das ebenso vorbildliche Know-how, Standards zu etablieren. Dies gelte allerdings nicht für den nun notwendigen Wissenstransfer und Know-how-Aufbau im Themenfeld Security.

KONTRAST ZU DEUTSCHLAND

„In den USA wird Security-Wissen offener ausgetauscht, an Standards mit Hochdruck gearbeitet", konstatierte Weimerskirch, der Gründer des Start-up Escrypt, der heutigen Etas-Tochter. Der Community-Gedanke sei in den USA stark ausgeprägt und das internationale Interesse sehr groß. Investoren seien allgegenwärtig. Dieser Dynamik steht Umfragen der Redaktion zufolge eine gewisse Verschlossenheit der Automobilhersteller gegenüber. Bei der Kommunikation von Security-Themen hat man sich herstellerübergreifend anscheinend auf ein Wording geeinigt. Wohl aufgrund dessen scheute man die Teilnahme an der ausgewogenen und ausführlichen Podiumsdiskussion. Dort herrschte Konsens, was den derzeitigen Status von IT-Konzepten und -Strategien in Zeiten eines intensiven Lernens und der Phase des Aufbaus betrifft. In puncto Security- Know-how schneiden deutsche OEMs auch gut ab, weiß Escrypt-Geschäftsführer Dr. Thomas Wollinger aus Kundenprojekten zu berichten. Die Ernsthaftigkeit, mit der Autohersteller unter anderem mit hohen Investitionen arbeiten, sei bemerkenswert: „Vor einigen Jahren war das noch nicht so". Auch in Erarbeitung der noch fehlenden Standards sei die Branche insbesondere in den vergangenen zwölf Monaten weitergekommen.

EHRLICHE EINSCHÄTZUNG

Der ehrlichen Einschätzung von Elmar Frickenstein in der Eröffnungs-Keynote folgte eine ebenso offene von Volvo-Entwicklungsvorstand Dr. Peter Mertens in seiner Abschluss-Keynote: Bis 2020 soll kein Mensch in einem neuen Volvo bei einem Unfall schwer verletzt oder bei einem Autounfall getötet werden, lautet eines seiner Zitate. Auf die Frage, wie sich denn das gesetzte Ziel erreichen ließe, antwortete der Ingenieur: „Das Ziel und die Vision ist wichtig und ernst gemeint. Ob wir es schaffen, wird die Zukunft zeigen." Und zum Hype des automatisierten Fahrens sagte er: „Ich halte viel davon, dass man darüber berichtet und spricht - und auch Pläne miteinander austauscht. Ich halte hingegen nichts davon, vollmundige Ankündigungen und Versprechungen zu machen, die dann nicht zu halten sind." Jeder verantwortliche Entwickler müsse sich an die Realisierung und Absicherung herantasten, beispielsweise mit einem Pilotprojekt mit Endkunden wie „DriveMe". Der Entwicklungschef betont: „Wir wissen, dass die bereits entwickelten Techniken zur Unfallvermeidung ziemlich gut sind. Sie bringen uns auf dem Weg zu Zero Fatalities entscheidend voran." Volvo sei an einem Punkt angekommen, selbst hochautomatisierte Fahrzeuge nicht mit Testingenieuren auf der Nordschleife des Nürburgrings, sondern im realen Straßenverkehr mit Endkunden zu testen.

[Quelle: ATZelektronik 11 (2016), Nr. 3, S. 70ff]

Druck:
Canon Deutschland Business Services GmbH
im Auftrag der KNV-Gruppe
Ferdinand-Jühlke-Str. 7
99095 Erfurt